Advanced Materials for Clinical Endodontic Applications

Advanced Materials for Clinical Endodontic Applications

Guest Editors

Saulius Drukteinis
Sivaprakash Rajasekharan
Matthias Widbiller

MDPI
Basel • Beijing • Wuhan • Barcelona • Belgrade • Novi Sad • Cluj • Manchester

Guest Editors
Saulius Drukteinis
Institute of Dentistry
Vilnius University
Vilnius
Lithuania

Sivaprakash Rajasekharan
Department of Paediatric
Dentistry
Ghent University
Ghent
Belgium

Matthias Widbiller
Department of Conservative
Dentistry and Periodontology
University Hospital Regensburg
Regensburg
Germany

Editorial Office
MDPI AG
Grosspeteranlage 5
4052 Basel, Switzerland

This is a reprint of the Special Issue, published open access by the journal *Journal of Functional Biomaterials* (ISSN 2079-4983), freely accessible at: https://www.mdpi.com/journal/jfb/special_issues/endodontic_mat.

For citation purposes, cite each article independently as indicated on the article page online and as indicated below:

Lastname, A.A.; Lastname, B.B. Article Title. *Journal Name* **Year**, *Volume Number*, Page Range.

ISBN 978-3-7258-5243-7 (Hbk)
ISBN 978-3-7258-5244-4 (PDF)
https://doi.org/10.3390/books978-3-7258-5244-4

Contents

About the Editors

Saulius Drukteinis

Saulius Drukteinis, a committed researcher and endodontics research group leader, boasts extensive expertise in clinical and experimental endodontic research and education. His expertise covers endodontic instruments, instrumentation techniques, and endodontic materials, particularly hydraulic calcium silicate-based formulations. His group conducts research integrating advanced imaging technologies, including micro-CT and SEM, to unravel microstructural features of novel endodontic instruments and materials. The Team's achievements include important insights into instrument properties, performance, and material characterization. The Team's research focuses on optimizing endodontic procedures through innovative instrument designs and material enhancements. His commitment to advancing the field is evident in his numerous peer-reviewed articles, textbook chapters, and international conference presentations, highlighting the practical applications of micro-CT and SEM in endodontic research.

Sivaprakash Rajasekharan

Sivaprakash Rajasekharan is a distinguished researcher and academic at Ghent University, Belgium. He brings a wealth of expertise in biomaterials and their applications in endodontics. His work spans clinical and experimental research, with a particular focus on the development and optimization of calcium silicate-based materials for regenerative and reparative dental therapies. Prof. Rajasekharan leads a dynamic research team dedicated to investigating the properties, performance, and clinical applications of novel biomaterials. Utilizing analytical techniques such as micro-CT, SEM, EDS, and FTIR spectroscopy, his group delves into the structural and functional characteristics of these materials, contributing significantly to the understanding of their behavior in clinical scenarios.

Under his guidance, his research team has made notable strides in improving the bioactivity and clinical efficacy of endodontic materials, emphasizing their role in minimally invasive dentistry and tissue regeneration. Prof. Rajasekharan's commitment to advancing dental materials science is reflected in his prolific contributions to high-impact journals, co-authored textbooks, and presentations at leading international conferences.

Matthias Widbiller

Matthias Widbiller, a dedicated clinician scientist and academic at the University of Regensburg, Germany, combines expertise in clinical endodontics with translational pulp biology research. He obtained his doctorate in dentistry in 2015 and subsequently joined the Department of Conservative Dentistry and Periodontology, where he is now a Full Professor of Endodontics and Head of the Research Laboratory. From 2017 to 2018, he expanded his scientific perspective as a postdoctoral fellow at the University of Texas Health Science Center at San Antonio in the renowned laboratory of Dr. Kenneth M. Hargreaves and Dr. Anibal Diogenes, focusing on tissue engineering of the dental pulp, bioactive materials, and dentin matrix proteins. His clinical expertise lies in vital pulp therapy, regenerative endodontic procedures, and dental traumatology, areas in which he has played a central role since establishing and coordinating the Centre for Dental Traumatology at the University Hospital of Regensburg in 2016. Prof. Widbiller's research group investigates cellular and molecular mechanisms underlying pulp regeneration, pulpal immune response, and the biocompatibility of innovative bioactive agents. His work aims to bridge fundamental biological insights with clinical applications, advancing strategies for preserving pulp vitality and improving patient outcomes.

Journal of
*Functional
Biomaterials*

MDPI

Editorial

Advanced Materials for Clinical Endodontic Applications: Current Status and Future Directions

Saulius Drukteinis [1,*], Sivaprakash Rajasekharan [2] and Matthias Widbiller [3]

1 Institute of Dentistry, Faculty of Medicine, Vilnius University, Zalgirio 115, LT-08217 Vilnius, Lithuania
2 Department of Paediatric Dentistry, School of Oral Health Sciences, Ghent University, B-9000 Ghent, Belgium; sivaprakash.rajasekharan@ugent.be
3 Department of Conservative Dentistry and Periodontology, University Hospital Regensburg, Franz-Josef-Strauß-Allee 11, D-93093 Regensburg, Germany; matthias.widbiller@ukr.de
* Correspondence: saulius.drukteinis@mf.vu.lt; Tel.: +370-610-41808

Endodontics has significantly evolved in recent years, with advancements in instruments, biomaterials and nanomaterials science playing a pivotal role [1,2]. These cutting-edge materials are transforming endodontic treatment techniques, offering enhanced properties, improved clinical outcomes and a more patient-centered approach [3]. Within the last decade, a wide variety of materials for root canal irrigation, disinfection and obturation, as well as for the management of endodontic complications, regenerative endodontic procedures (REP), endodontic surgery and pediatric endodontic treatment, have been introduced. Therefore, detailed in vitro, in vivo and clinical investigations of these materials are crucial for their scientifically based, standardized, safe and successful use in daily clinical practice [4].

The quality of root canal obturation and materials used for this purpose, including endodontic sealers, play a significant role in the success of endodontic treatments. These materials should ensure a three-dimensional seal within the root canal system, preventing re-infection and ensuring the longevity of the treatment and the survival of the endodontically treated tooth. Over the years, various endodontic sealers have been developed, each with unique properties, advantages and disadvantages [5]. The choice of sealer depends on various factors, including the specific clinical scenario, the clinician's experience and preference and the desired properties of the material. Regarding the latest advancements, hydraulic calcium silicate-based (HCS) sealers, commonly called "bioceramic sealers", have gained tremendous popularity in endodontics [6–8]. These sealers, especially the fourth and fifth generations, are being intensively investigated due to their relatively recent introduction to the market, their modified chemical composition, their advanced physical and biological properties and the different behavior of these materials in a biological environment [9,10]. All these changes have greatly simplified their clinical applications, even for operators with limited clinical experience [11,12].

For decades, clinicians have successfully used conventional cold or warm (thermoplastic) compaction root canal obturation techniques with a favorable prognosis for endodontically treated teeth [13]. The fundamental principle of these techniques is increasing the gutta-percha volume and minimizing the sealer amount [14]. The development of HCS sealers has significantly changed these principles of root canal obturation. Meanwhile, due to a lack of shrinkage and long-term dimensional stability, these materials can be used in larger volumes for sealer- or filler-based obturations without increasing the amount of gutta-percha in the root canal. Although all obturation techniques are equally effective, the single-cone (SC) obturation technique is easier to apply, especially for inexperienced clinicians [15]. Moreover, the scientific background indicates that the biocompatible, bioactive and antibacterial HCS materials that slightly expand upon setting and remain dimensionally stable in conjunction with a simplified SC obturation can provide much better results than lateral condensation and surpass it as the most efficient endodontic sealing method [16,17].

Endodontic treatment and retreatment, while highly successful, can occasionally present possible complications that require specialized and very complex management [18]. Likewise, specific clinical scenarios may necessitate endodontic surgery to achieve desirable treatment outcomes [19]. Techniques for managing endodontic complications and endodontic surgery have changed and significantly require advanced endodontic materials [11]. Nowadays, HCS materials are considered superior for the management of endodontic complications due to their biological properties, such as biocompatibility, bioactivity and osteogenic potential [10,20]. They are also hydrophilic and hygroscopic, and as they require moisture to set, they are compatible with wet environments like live pulp or periapical tissues [21]. Many materials are currently available for managing endodontic complications, including flowable materials launched as premixed and ready-to-use pastes or powder/liquid formulations [11]. The fourth and fifth types of HCS materials are the most popular in modern endodontics. Both types are tri-calcium silicate-based; however, type four materials are mixed with water, while type five are pre-mixed and ready-to-use [10]. Some materials are only suggested to be used as dentin or root repair materials in conjunction with different application techniques. In contrast, other materials are proposed as sealers or biological fillers and can be used for root canal obturation and root repair. The main advantages of flowable HCS materials are their ease of manipulation and clinical applicability [9].

However, despite all these advantages, they present possible drawbacks. The setting of these materials is based on a hydration reaction, for which an appropriate level of moisture is required [21]. The fourth generation of HCS sealers are water-based materials, and they receive moisture directly from the liquid they are mixed with. Meanwhile, the fifth generation, or premixed formulations, requires environmental moisture [21]. To ensure the complete hydration-based setting of these materials, the root canal should not be overdried [9,21]. Ensuring that the canal is sufficiently dry can be challenging for clinicians, as it is tricky to determine how dry the root canal should be in every clinical case to ensure the complete setting of the premixed HCS. Therefore, further investigations are needed to determine the optimal amount of moisture needed for the materials to set and translate these findings into clinical protocols. Another disadvantage of HCS materials is that water-based formulations are sensitive to heat, and their application when using thermoplastic obturation techniques can significantly change their properties [5]. Therefore, the fourth generation of HCS materials should be used only with cold obturation techniques, while these restrictions are not relevant for premixed formulations [12].

The proper selection of root canal irrigants and efficient irrigation techniques are essential when cleaning and disinfecting the root canal system [22]. Biofilms within the anatomic complexities of the root canal system contain highly resistant bacteria, which are difficult to eradicate with currently available irrigants [22]. Sodium hypochlorite has remained the most widely used irrigant for decades. However, chelators should supplement it to enhance the effectiveness of the irrigation [23], and mixtures of different active substances and nanoparticles have additionally been trialed for this purpose [23]. The delivery of the irrigants using a syringe and needle and their sonic or ultrasonic activation are currently the most popular irrigation methods. At the same time, different agitation and activation methods using negative and positive pressure, or lasers, are widely used and have been investigated to enhance the removal of debris and biofilm from root canal irregularities [24]. Unfortunately, despite these innovations, there is no solid scientific evidence that any adjunct irrigation method can remarkably improve the long-term outcome of root canal treatment [4]. Therefore, redefining this field's research priorities and strategies, focusing on multidisciplinary scientific approaches and clinically relevant comparisons, is necessary [22].

Some researchers argue that a temporary antibacterial root canal dressing material should be used to enhance root canal disinfection during the endodontic treatment of teeth with pulp necrosis and apical periodontitis, as well as for the management of endodontic complications or regenerative endodontic procedures (REPs) [25]. For decades, calcium hydroxide was the material used for these purposes. Recently, ready-to-use bioceramic-based pastes have been introduced, such as BIO-C® TEMP (Angelus, Londrina, Brazil) and EndoSequence BC Temp (Brasseler USA, Savannah, GA, USA), for intracanal dressing [26]. These materials have antimicrobial properties and are recommended as a substitute for conventional calcium hydroxide dressings, as they are easily washed out of the root canals, eliminating the need for additional irrigation with citric acid, unlike conventional calcium hydroxide paste, which is difficult to remove from the root canal system [27].

Endodontic surgery is indicated when non-surgical endodontic retreatment fails or is impossible. Due to the rapid developments in implant dentistry, endodontic surgery has become a less popular treatment option than tooth replacement with an implant. However, well-designed clinical trials have shown that endodontic retreatment and surgical endodontic procedures are equally effective, if not superior, treatment options [28]. Recently, the fourth and fifth generations of HCS root repair materials have also become increasingly popular for use in endodontic surgery [9,11]. The current scientific findings indicate that these materials have superior properties and handling characteristics and provide healing rates after endodontic surgery similar to those of previously used MTA cement [29,30]. These commercially available plasticized putty-type repair materials are identical in their chemical composition, physical and biological properties and are clinically equally effective [31].

Regenerative endodontics represents a transformative shift in the field of endodontics, focusing on the regeneration of dental pulp tissues and the restoration of tooth vitality [32]. These new biology-based and less invasive concepts have entered the clinical routine, expanding the principles of classical endodontics. REPs, as a biologically based approach, utilize tissue engineering to replace damaged structures, including dentin and cells of the pulp–dentin complex [33]. These procedures aim to heal apical lesions, resolve clinical signs and symptoms, continue root development and strengthen dentin tissue to prevent potential root fractures [25], with the aim of ultimately regenerating a functional pulp–dentin complex and restoring regular nociception. Although stem cells, growth factors, scaffolds and proper disinfection are essential to the overall success of REPs, the biocompatible materials used to cover blood clots play a critical role in the outcome of the treatment [34]. Nowadays, fourth- and fifth-generation HCS materials are recommended for REPs [35].

Pediatric endodontics presents different challenges than adult endodontics, given pediatric patients' unique anatomical, physiological and behavioral characteristics. The use of appropriate endodontic instruments, materials and techniques in pediatric endodontic cases is crucial to ensuring effective and less time-consuming treatment, patient comfort and a favorable long-term prognosis. Recent material advances in pediatric endodontics address these specific requirements, and a more conservative and minimally invasive approach is usually preferred, particularly vital pulp therapies that preserve dental pulp vitality whenever possible. Nonvital pulp therapy in primary teeth has considerable limitations and is only considered if the pulp is inflamed irreversibly, necrotic and infected, while the roots of the teeth should have minimal or no signs of resorption. The rotary files for pediatric cases were designed for faster root canal shaping, while resorbable injectable filling pastes were proposed for obturation [36]. However, in primary teeth, priority should always be given, whenever possible, to vital pulp therapy based on pulpotomy procedures, including when using hydraulic calcium silicate-based materials [37].

Therefore, this Special Issue, "Advanced Materials for Clinical Endodontic Applications", aims to deepen researchers' understanding of the challenges surrounding advanced endodontic materials. However, the current scientific data indicate that there is a critical need to focus on interdisciplinary research that bridges the gap between material science and clinical endodontics. We should translate recent research findings into clinical solutions and protocols to improve endodontic treatment standards and outcomes.

Author Contributions: Conceptualization, S.D., S.R. and M.W.; writing—original draft preparation, S.D., S.R. and M.W.; writing—review and editing, S.D., S.R. and M.W. All authors have read and agreed to the published version of the manuscript.

Conflicts of Interest: The authors declare no conflicts of interest.

References

1. Zanza, A.; D'angelo, M.; Reda, R.; Gambarini, G.; Testarelli, L.; Di Nardo, D. An Update on Nickel-Titanium Rotary Instruments in Endodontics: Mechanical Characteristics, Testing and Future Perspective—An Overview. *Bioengineering* **2021**, *8*, 218. [CrossRef]
2. Guerrero-Gironés, J.; Forner, L.; Luis Sanz, J.; Rodríguez-Lozano, F.J.; Ghilotti, J.; Llena, C.; Lozano, A.; Melo, M. Scientific production on silicate-based endodontic materials: Evolution and current state: A bibliometric analysis. *Clin. Oral Investig.* **2002**, *26*, 5611–5624. [CrossRef]
3. Duncan, H.F.; Nagendrababu, V.; El-Karim, I.A.; Dummer, P.M.H. Outcome measures to assess the effectiveness of endodontic treatment for pulpitis and apical periodontitis for use in the development of European Society of Endodontology (ESE) S3 level clinical practice guidelines: A protocol. *Int. Endod. J.* **2021**, *54*, 646–654. [CrossRef]
4. Duncan, H.F.; Kirkevang, L.L.; Peters, O.A.; El-Karim, I.; Krastl, G.; Del Fabbro, M.; Chong, B.S.; Galler, K.M.; Segura-Egea, J.J.; Kebschull, M. Treatment of pulpal and apical disease: The European Society of Endodontology (ESE) S3-level clinical practice guideline. *Int. Endod. J.* **2023**, *56* (Suppl. S3), 238–295. [CrossRef] [PubMed]
5. Pirani, C.; Camilleri, J. Effectiveness of root canal filling materials and techniques for treatment of apical periodontitis: A systematic review. *Int. Endod. J.* **2023**, *56* (Suppl. S3), 436–454. [CrossRef]
6. Prati, C.; Gandolfi, M.G. Calcium silicate bioactive cements: Biological perspectives and clinical applications. *Dent. Mat.* **2015**, *31*, 351–370. [CrossRef]
7. Camilleri, J. Characterization and Properties of Bioceramic Materials for Endodontics. In *Bioceramic Materials in Clinical Endodontics*, 1st ed.; Drukteinis, S., Camilleri, J., Eds.; Spinger: Cham, Germany, 2021; pp. 7–18.
8. Camilleri, J. Current Classification of Bioceramic Materials in Endodontics. In *Bioceramic Materials in Clinical Endodontics*, 1st ed.; Drukteinis, S., Camilleri, J., Eds.; Spinger: Cham, Germany, 2021; pp. 1–6.
9. Sfeir, G.; Zogheib, C.; Patel, S.; Giraud, T.; Nagendrababu, V.; Bukiet, F. Calcium Silicate-Based Root Canal Sealers: A Narrative Review and Clinical Perspectives. *Materials* **2021**, *14*, 3965. [CrossRef] [PubMed]
10. Camilleri, J.; Atmeh, A.; Li, X.; Meschi, N. Present status and future directions: Hydraulic materials for endodontic use. *Int. Endod. J.* **2022**, *55* (Suppl. S3), 710–777. [CrossRef] [PubMed]
11. Dawood, A.E.; Parashos, P.; Wong, R.H.K.; Reynolds, E.C.; Manton, D.J. Calcium silicate-based cements: Composition, properties, and clinical applications. *J. Investig. Clin. Dent.* **2017**, *8*, e12195. [CrossRef]
12. Trope, M.; Bunes, A.; Debelian, G. Root filling materials and techniques: Bioceramics a new hope? *Endod. Top.* **2015**, *32*, 86–96. [CrossRef]
13. Whitworth, J. Methods of filling root canals: Principles and practices. *Endod. Top.* **2005**, *12*, 2–24. [CrossRef]
14. Camilleri, J. Will Bioceramics be the Future Root Canal Filling Materials? *Curr. Oral Health Rep.* **2017**, *4*, 228–238. [CrossRef]
15. Drukteinis, S.; Bilvinaite, G.; Tusas, P.; Shemesh, H.; Peciuliene, V. Porosity Distribution in Single Cone Root Canal Fillings Performed by Operators with Different Clinical Experience: A microCT Assessment. *J. Clin. Med.* **2021**, *10*, 2569. [CrossRef] [PubMed]
16. Penha Da Silva, P.J.; Marceliano-Alves, M.F.; Provenzano, J.C.; Dellazari, R.L.A.; Gonçalves, L.S.; Alves, F.R.F. Quality of Root Canal Filling Using a Bioceramic Sealer in Oval Canals: A Three-Dimensional Analysis. *Eur. J. Dent.* **2021**, *15*, 475–480. [CrossRef] [PubMed]
17. Angerame, D.; De Biasi, M.; Pecci, R.; Bedini, R. Filling ability of three variants of the single-cone technique with bioceramic sealer: A micro-computed tomography study. *J. Mater. Sci. Mater. Med.* **2020**, *31*, 91. [CrossRef] [PubMed]
18. Siew, K.; Lee, A.H.C.; Cheung, G.S.P. Treatment outcome of repaired root perforation: A systematic review and meta-analysis. *J. Endod.* **2015**, *41*, 1795–1804. [CrossRef] [PubMed]
19. Floratos, S.; Kim, S. Modern Endodontic Microsurgery Concepts: A Clinical Update. *Dent. Clin. N. Am.* **2017**, *61*, 81–91. [CrossRef] [PubMed]
20. Wang, Z. Bioceramic materials in endodontics. *Endod. Top.* **2015**, *32*, 3–30. [CrossRef]
21. Guo, Y.; Du, T.; Li, H.; Shen, Y.; Mobuchon, C.; Hieawy, A.; Wang, Z.; Yang, Y.; Jingzhi, M.J.; Haapasalo, M. Physical properties and hydration behavior of a fast-setting bioeramic endodontic material. *BMC Oral Health* **2016**, *16*, 23. [CrossRef]
22. Boutsioukis, C.; Arias-Moliz, M.T. Present status and future directions—Irrigants and irrigation methods. *Int. Endod. J.* **2022**, *55* (Suppl. S3), 588–612. [CrossRef]
23. Fernandes Zancan, R.; Hadis, M.; Burgess, D.; Zhang, Z.J.; Maio, A.D.; Tomson, P.; Duarte, M.A.H.; Camilleri, J. A matched irrigation and obturation strategy for root canal therapy. *Sci. Rep.* **2021**, *11*, 4666. [CrossRef]
24. Ali, A.; Bhosale, A.; Pawar, S.; Kakti, A.; Bichpuriya, A.; Agwan, M.A. Current Trends in Root Canal Irrigation. *Cureus* **2022**, *14*, 24833. [CrossRef]
25. Wei, X.; Yang, M.; Yue, L.; Huang, D.; Zhou, X.; Wang, X.; Zhang, Q.; Qiu, L.; Huang, Z.; Wang, H.; et al. Expert consensus on regenerative endodontic procedures. *Int. J. Oral Sci.* **2022**, *14*, 55. [CrossRef]

26. Guerreiro, J.C.M.; Ochoa-Rodrigez, V.M.; Rodrigues, E.M.; Chavez-Andrade, G.M.; Tanomaru-Filho, M.; Guerreiro-Tanomaru, J.M.; Faria, G. Antibacterial activity, cytocompatibility and effect of Bio-C Temp bioceramic intracanal medicament on osteoblast biology. *Int. Endod. J.* **2021**, *54*, 1155–1165. [CrossRef]
27. Topçuoğlu, H.S.; Düzgün, S.; Ceyhanlı, K.T.; Aktı, A.; Pala, K.; Kesim, B. Efficacy of different irrigation techniques in the removal of calcium hydroxide from a simulated internal root resorption cavity. *Int. Endod. J.* **2015**, *48*, 309–316. [CrossRef]
28. Huang, S.; Chen, N.N.; Yu, V.S.H.; Lim, H.A.; Lui, J.N. Long-term Success and Survival of Endodontic Microsurgery. *J. Endod.* **2020**, *46*, 149–157. [CrossRef] [PubMed]
29. Pinto, D.; Marques, A.; Pereira, J.F.; Palma, P.J.; Santos, J.M. Long-Term Prognosis of Endodontic Microsurgery—A Systematic Review and Meta-Analysis. *Medicina* **2020**, *56*, 447. [CrossRef] [PubMed]
30. Koutroulis, A.; Valen, H.; Ørstavik, D.; Kapralos, V.; Camilleri, J.; Sunde, P.T. Effect of exposure conditions on chemical properties of materials for surgical endodontic procedures. *Eur. J. Oral Sci.* **2023**, *131*, 12943. [CrossRef] [PubMed]
31. Primus, C.; Gutmann, J.L.; Tay, F.R.; Fuks, A.B. Calcium silicate and calcium aluminate cements for dentistry reviewed. *J. Am. Ceram. Soc.* **2022**, *105*, 1841–1863. [CrossRef]
32. Hargreaves, K.M.; Diogenes, A.; Teixeira, F.B. Treatment Options: Biological Basis of Regenerative Endodontic Procedures. *J. Endod.* **2013**, *39*, 30–43. [CrossRef] [PubMed]
33. Galler, K.M. Clinical procedures for revitalization: Current knowledge and considerations. *Int. Endod. J.* **2016**, *49*, 926–936. [CrossRef] [PubMed]
34. Staffoli, S.; Plotino, G.; Nunez Torrijos, B.G.; Grande, N.M.; Bossù, M.; Gambarini, G.; Polimeni, A. Regenerative Endodontic Procedures Using Contemporary Endodontic Materials. *Materials* **2019**, *12*, 908. [CrossRef] [PubMed]
35. Galler, K.M.; Widbiller, M.; Camilleri, J. Bioceramic materials in regenerative endodontics. In *Bioceramic Materials in Clinical Endodontics*, 1st ed.; Drukteinis, S., Camilleri, J., Eds.; Spinger: Cham, Germany, 2021; pp. 29–38.
36. Jindal, L.; Bhat, N.; Mehta, S.; Bansal, S.; Sharma, S.; Kumar, A. Rotary Endodontics in Pediatric Dentistry: Literature Review. *Int. J. Health Biol. Sci.* **2020**, *3*, 9–13.
37. Primus, C. Bioactive Ceramics for Pediatric Dentistry. In *Contemporary Endodontics for Children and Adolescents*, 1st ed.; Fuks, A.B., Moskovitz, M., Tickotsky, N., Eds.; Spinger: Cham, Germany, 2023; pp. 149–187.

Journal of
Functional
Biomaterials

MDPI

Review

Engineering Dental Tissues Using Biomaterials with Piezoelectric Effect: Current Progress and Future Perspectives

Sumanta Ghosh [1], Wei Qiao [1], Zhengbao Yang [2], Santiago Orrego [3,4] and Prasanna Neelakantan [1,*]

[1] Faculty of Dentistry, The University of Hong Kong, Hong Kong SAR, China
[2] Department of Mechanical Engineering & Department of Materials Science and Engineering, City University of Hong Kong, Hong Kong, China
[3] Oral Health Sciences Department, Kornberg School of Dentistry, Temple University, Philadelphia, PA 19140, USA
[4] Bioengineering Department, College of Engineering, Temple University, Philadelphia, PA 19140, USA
* Correspondence: prasanna@hku.hk

Abstract: Dental caries and traumatic injuries to teeth may cause irreversible inflammation and eventual death of the dental pulp. Nevertheless, predictably, repair and regeneration of the dentin-pulp complex remain a formidable challenge. In recent years, smart multifunctional materials with antimicrobial, anti-inflammatory, and pro-regenerative properties have emerged as promising approaches to meet this critical clinical need. As a unique class of smart materials, piezoelectric materials have an unprecedented advantage over other stimuli-responsive materials due to their inherent capability to generate electric charges, which have been shown to facilitate both antimicrobial action and tissue regeneration. Nonetheless, studies on piezoelectric biomaterials in the repair and regeneration of the dentin-pulp complex remain limited. In this review, we summarize the biomedical applications of piezoelectric biomaterials in dental applications and elucidate the underlying molecular mechanisms contributing to the biological effect of piezoelectricity. Moreover, we highlight how this state-of-the-art can be further exploited in the future for dental tissue engineering.

Keywords: bio-piezoelectricity; dentin-pulp complex; mechanosensitive ion channel; piezo-1 receptor; piezoelectric biomaterials; regenerative endodontics; tissue engineering

1. Introduction

The dental pulp is a vital tissue that is enclosed within the highly mineralized dentin. Injury to the dental pulp due to microbial infection (caries) and traumatic injuries to the tooth result in its inflammation and subsequent necrosis if left untreated. More than 25% of school children and 33% of adults experience trauma to their permanent teeth before 19 years of age [1]. While root canal treatment is predictable and highly successful in mature teeth, the same is not true for immature teeth because of thin and weak dentinal walls, which predispose them to fracture under stress overload [2]. Regeneration of the dentin-pulp complex is a preferred treatment for such cases, and this procedure aims to engineer metabolically active pulp or pulp-like tissue that can form new dentin (i.e., dentin-pulp complex), prevent re-infection of the tooth, and improve the functionality of the tooth [1,2]. However, this remains a clinical challenge given the lack of multifunctional materials that can control infection and simultaneously support the regeneration of tissues.

Physiological load in tissues such as the skin, bone, cartilage, periodontal ligament, and dentin results in the activation of specific molecular cell signaling processes that trigger the generation of electrical potential, a property termed piezoelectricity. The generation of electric potential results in depolarization (or the potential difference between the cell surface and the cytoplasm), inducing various cell signaling pathways such as the calcium/calmodulin pathway in bone, cartilage, and tendon [3,4]. The resultant electric potential depolarizes the cell membrane and leads to the opening of several voltage-gated

Ca^{2+} ion channels across the cell membranes. Then, a rapid influx of Ca^{2+} ions result in increased intracellular Ca^{2+} ions concentrations, which causes the dephosphorylation of nuclear factors of activated cells (NF-AT). Subsequently, the translocation of NF-AT into the nucleus regulates the transcription of several genes, which results in several cellular functions, including cell proliferation and stem cell differentiation (Figure 1). The piezoelectric potential also induces phosphatidylinositol 3-kinase (PI3K)/protein kinase B (Akt) based signaling, which regulates the fundamental aspects of physiological wound healing processes, including cell proliferation, differentiation, and migration, angiogenesis, and metabolism [4,5]. Additionally, this pathway also stimulates the extracellular signal-regulated kinase (ERK) and GTPase-mediated actin polymerization process, which ultimately leads to cutaneous wound healing by promoting keratinocyte proliferation and migration [5]. Upregulation of TGF-β, BMP, and COL-III by piezoelectric mechanisms is crucial for bone and cartilage tissue repair or regeneration [6].

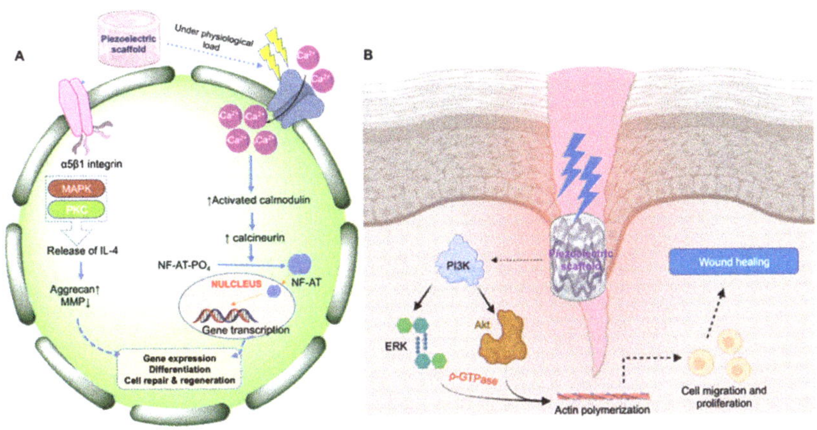

Figure 1. Schematic representation of cell signaling pathways through which piezoelectric materials elicit their action in tissue repair and regeneration. (**A**) Calcium/calmodulin pathway in bone, cartilage, and tendon; (**B**) PI3K-Akt-based signaling pathway for wound healing in skin.

Furthermore, to elucidate the cellular and molecular level mechanisms of the piezoelectric potential generated from the piezo materials, many gene/protein biomarkers have been identified for different tissue regeneration over the past few years. For instance, Bhang et al. [5] identified that several intra/extra-cellular proteins such as transforming growth factor (TGF-β), vascular endothelial growth factor (VEGF), collagen type-3 (Col III), integrin α5, proliferating cell nuclear antigen (PCNA), keratin 17, MMP2, CD68, and fibronectin are expressed after the induction of the piezoelectric potential on human dermal fibroblasts (HDFs), keratinocytes, and hMSCs, which directly indicate accelerated inflammation modulation, re-epithelialization, proliferation, granulation, and remodeling stages of the wound healing process. Additionally, Vignesh et al. reported the upregulation of anti-runt-related transcription factor 2 (RUNX2), osteocalcin (OCN), and osteopontin (OPN) gene expression, which enhances alkaline phosphatase and calcium deposition, indicating the osteogenetic differentiation ability of piezoelectric biomaterials for bone regeneration [3,7,8]. Nevertheless, it has also been reported that the piezoelectric potential positively influences the upregulation of several neurotropic factors such as NGF, GDNF, and BDNF, which are required for axonal regeneration and neuronal differentiation for nerve tissue engineering [9,10]. Thus, different genes or biomarkers involved in the different physiological activities of the piezoelectric potential are summarized in Table 1.

Predictable regeneration of the dentin-pulp complex requires a material to be endowed with multiple properties, including antimicrobial, immune modulatory, and cell differentiation to several phenotypes (odontogenic, vascular, and neural) [11–14]. Given that biological tissues such as dentin respond to mechanical stimulus by producing piezoelectric potential, there is a remarkable interest in developing mechano-responsive, piezoelectric biomaterial-based scaffolds for the repair and regeneration of dental tissues [11–13]. Indeed, such materials have shown tissue differentiation, antimicrobial, and anti-inflammatory properties. In this review, we first describe the basic principles and mechanisms of piezoelectricity. We then report a detailed bibliometric analysis of the published research on piezoelectric biomaterials used for tissue engineering. Then, we critically discuss promising piezoelectric biomaterials that could be exploited for engineering the dentin-pulp complex and the methods by which piezoelectric scaffolds may be fabricated.

Table 1. Biomarkers involved in the regenerative roles of piezoelectric biomaterials in various tissues.

Piezoelectric Biomarker	Role in Tissue Regeneration	Reference
Increased expression of transforming growth factor (TGF-β)	Voltage-gated Ca^{2+} ion channel opening	[15]
Increased expression of bone morphogenetic protein (BMP)	Bone remodeling	[15]
Increased expression of collagen type-3 (Col III)	Collagen synthesis and tissue granulation	[15]
Increased expression of collagen type 4 (Col IV)	Keratinocyte migration	[15]
Upregulation of CD68	Macrophage differentiation, anti-inflammatory	[16]
Upregulation of vascular endothelial growth factor (VEGF)	Anti-inflammatory	[15]
Upregulation of integrin α5	Angiogenesis	[15]
Upregulation of CD99	Angiogenesis	[15]
Extracellular signal-regulated protein kinase (ERK1/2)	Electrostatic migration	
Proliferating cell nuclear antigen (PCNA)	Cell proliferation and signaling	[15]
α-actin	Myofibroblastic differentiation	[15]
Rho-GTPase	Electrotaxis	[17]
Enhanced phosphorylation of PI3K	Electrotaxis	[17]
Enhanced phosphorylation of Akt	Electrotaxis	[18]
Downregulation of Scleraxis	Tenogenic differentiation	[19]
Runt-related transcription factor 2 (Runx 2)	Osteogenic differentiation	[20]
Myoblast determination protein (MyoD)	Terminal myogenic differentiation	[21]
Myocyte enhancer factor-2 (MEF2)	Myogenic differentiation	[21]

2. Basic Principles of the Piezoelectric Effect

The term 'piezoelectric' is derived from the Greek word 'piezein' meaning pressure. The preliminary hypothesis of the 'piezoelectric effect' was proposed by the renowned French physicists Jacques and Pierre Curie in 1880 [22]. However, it was not until 1946 that Cady [23] deciphered the underlying principle of the piezoelectric properties of barium titanate ($BaTiO_3$, BT) and is recognized as the 'Father of Piezoelectricity' [24–26].

The basic principle of the piezoelectric effect is the crystal habit deformation of a material under definite mechanical stress [26], which is attributed to their asymmetric crystal architecture and microcrystalline hierarchy. Typically, any material can experience mechanical stimuli from the external environment. This could be elongation or tension, twisting or shear, bending or torsional, and squeezing or compression forces, depending on the direction of the mechanical loads. Mechanical-stress-induced deformation causes a relative shifting of the positive and negative charge center in the material crystal architecture, generating the motion of an electric dipole or polarization. This aligned distorted electric dipole state leads to the generation of electrical potential and causes a charge flow. In contrast, in non-piezoelectric materials, the overall charge center of positive and negative ions in the unit cell coincides, and even with applied deformation, these electric dipoles cancel out, and no overall polarization appears (Figure 2).

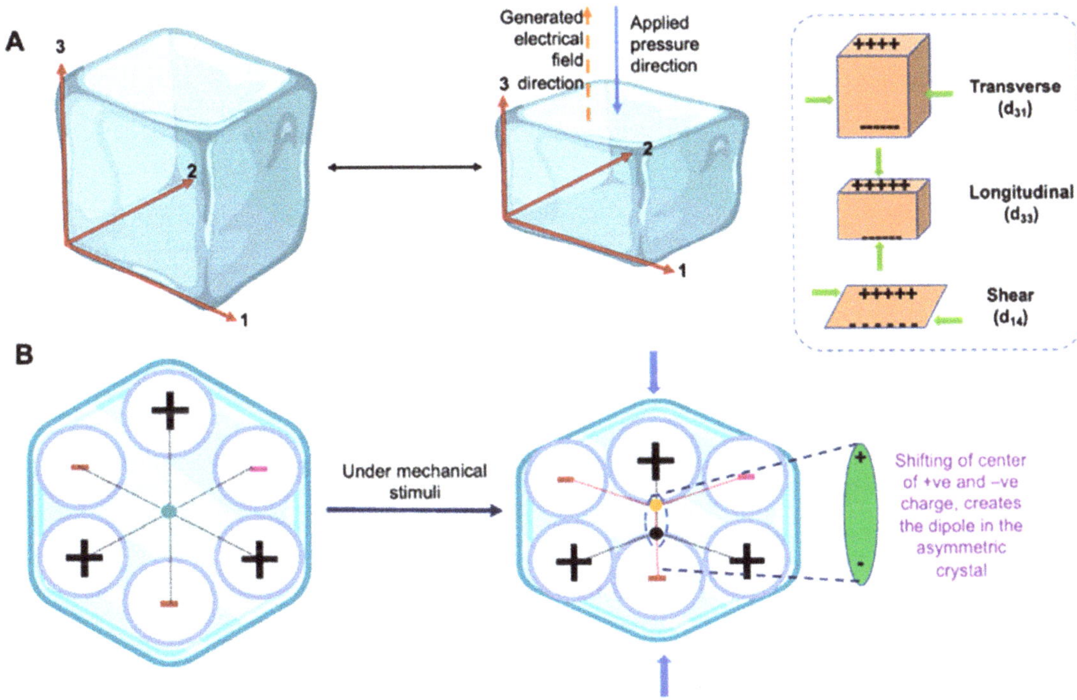

Figure 2. Fundamentals of the piezoelectric effect. (**A**) The direction of the piezoelectric coefficient, i.e., d_{33}, implies that the generation of electric potential takes place in direction 3 in response to the applied stress from direction 3; (**B**) Mechanism of piezoelectric potential generation in a crystal.

Piezoelectricity is a linear and reversible process. Transformation of a mechanical stimulus into electrical potential is termed the 'direct piezoelectric effect', whereas the reverse phenomenon is known as the 'indirect piezoelectric effect'. Real-world applications of piezoelectric materials were limited to acoustic devices based on the 'indirect piezoelectric effect' in several healthcare devices, including ultrasound transducers and MRI contrast agents [27], until 'direct piezoelectricity' was demonstrated in quartz and Rochelle salt. It was further discovered that crystalline polymorphic materials such as barium titanate ($BaTiO_3$), lithium niobate ($LiNbO_3$), calcium titanate ($CaTiO_3$), and strontium titanate ($SrTiO_3$) exhibited a direct piezoelectric effect due to phase transition from one crystal structure to another, a phenomenon known as the 'ageing effect' [28]. For instance, barium titanate (BT) crystal shows phase transition from a non-piezoelectric cubic structure to a non-centrosymmetric tetragonal structure under mechanical stress. As piezoelectricity is an inherent physical property of the material (such as pyroelectricity and ferroelectricity), it loses its effect after heating past the 'Curie temperature' (T_c). Different materials exhibit different Tc, which limits, in some cases, the real-world applications. For example, T_c for BT is 120 °C whereas for $LiNbO_3$, it is 1140 °C, making the latter suitable for high-energy applications such as energy harvesting devices and tire pressure monitoring systems [29].

The piezoelectric coefficient (or piezoelectric modulus) is a material constant that relates the polarization per unit of the electric field experienced by the material per unit of the applied mechanical stress and is expressed in the unit of Coulomb/Newton (C/N). This coefficient is expressed as 'd_{ij}', where 'i' indicates the direction of polarization in the material and 'j' indicates the direction of the applied stress (or induced strain). For example, in the constant d_{33}, the generation of electric potential takes place in direction

3 in response to the applied stress from direction 3 (Figure 2), which means that the mechanical load is applied parallel to the polarization axis [27,28]. The piezoelectric charge coefficient is most frequently used to evaluate the goodness of a piezoelectric material. The piezoelectric coefficient has different magnitudes depending on the direction of the applied mechanical stress and polarization. In the case of d_{31}, the charge is collected from the same surface as d_{33}, but the force is applied at right angles to the polarization axis, rending a different constant value. For example, for BT, d_{33} is 90–788 pC/N whereas d_{31} is −33.4 to −78 pC/N [29].

The interactions of piezoelectric materials with biological processes in a mammalian cell is termed 'bio-piezoelectricity' [7,15]. Piezoelectric charges play a role in various physiological processes, including cell division, migration, differentiation, and regeneration. Notably, the effect of bio-piezoelectricity on cells is more prominent in tissues that undergo normal physiological movement or in stress-bearing organs [2,15] due to the proportional relationship between the piezoelectric potential and the applied stress. For example, tissues such as bone and cartilage are dynamically stimulated by functional loads, and piezoelectric scaffolds may stimulate the regenerative signaling pathways to enhance tissue regeneration at the impaired site [8]. However, the interaction between the dentin-pulp complex and piezoelectric biomaterials remains poorly investigated and is an important avenue for future research.

3. Bibliometric Analysis of Research on Piezoelectric Biomaterials for Tissue Engineering

To inform the status of piezoelectric-based biomaterials for tissue engineering, we performed a bibliometric analysis using data acquired from three databases (PubMed, Scopus, and SciFinder) based on relevant keywords ('piezoelectric biomaterials', 'piezoelectric polymers', 'tissue engineering', 'bone, cartilage, neural, skin, and dental'). We then excluded reviews, systematic reviews, hypotheses, and literature-survey-based articles to ensure that only original research articles were included. To focus our sample space of the analysis on the direct applications of piezoelectric biomaterials on tissue engineering, we excluded research articles related to the topic of piezoelectric biosensors, implantable sensors, and drug delivery.

A total of 271 articles published from 1982 to 2021 formed the sample space. It was observed that from 1982 to 2021, 58% of the work on piezoelectric biomaterials was performed on bone, cartilage, and tendon tissue engineering while 13% and 10% of the work was performed on skin and neuronal tissue engineering, respectively. Notably, only 3% of the work was associated with dental tissue engineering (Figure 3). There was a continuous increase in the number of published research articles from 1982 to 2011 in piezoelectricity-based dental tissue engineering. We then used VOSviewer software to correlate the bibliometric data and visualize the interlinks between different piezo scaffolds with their biological properties. Most of the piezoelectric biomaterials were fabricated into three most preferred forms, i.e., nanofibers, hydrogels, and 3D-printed scaffolds composed of polymeric materials to support tissue repair and regeneration. In general, it was observed that studies on the application of piezoelectric biomaterials in engineering to a range of hard and soft tissues, including bone, cartilage, skin, and dental tissues such as pulp, periodontal ligament, and alveolar bone were closely linked to studying cellular responses such as adhesion, differentiation, and proliferation. Characterization of the mechanical strength and the ensuing mechanisms were explored for piezoelectric dental biomaterials.

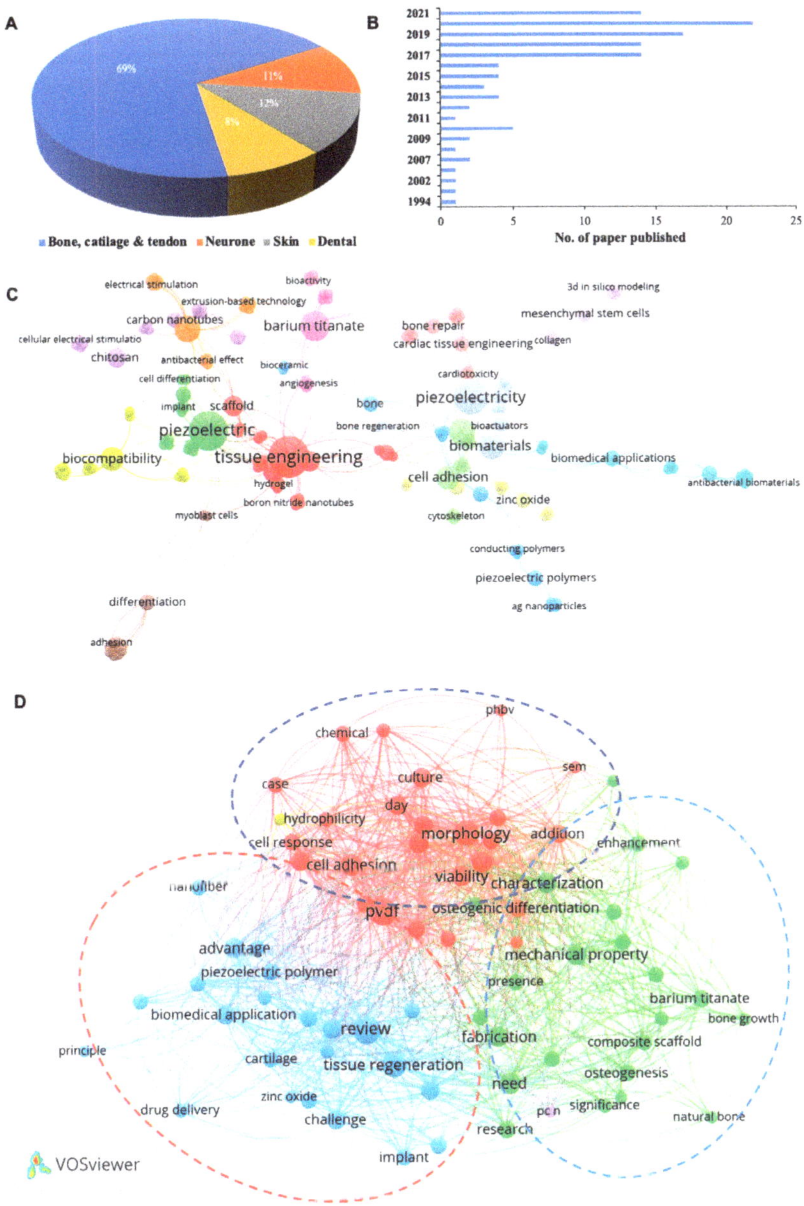

Figure 3. *Cont.*

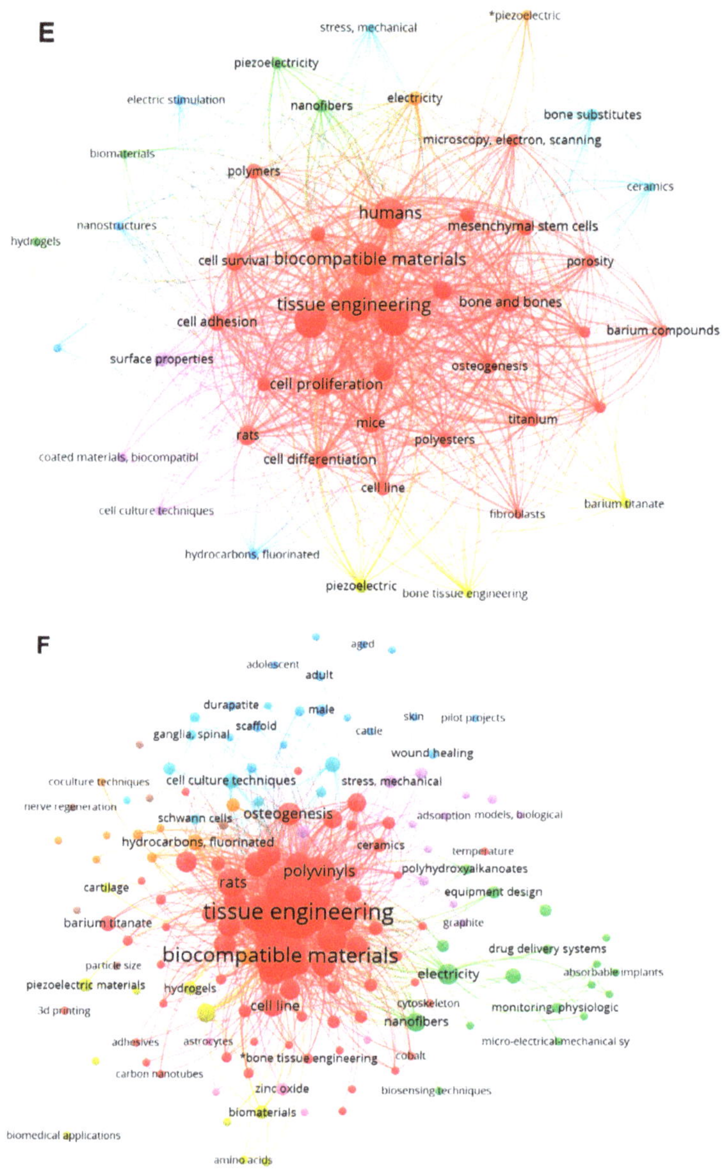

Figure 3. Bibliographic analysis of piezoelectric biomaterials for tissue engineering applications; (**A**) Pie chart of the percentage of articles published on tissue engineering applications of piezo-materials; (**B**) Year-wise number of studies published on piezoelectric biomaterials; (**C**) Bibliographic coupling analysis of piezoelectric biomaterials for different tissue engineering applications; (**D–F**) Bibliographic coupling analysis for piezoelectric scaffold for bone and cartilage tissue engineering and skin tissue, respectively (performed and schematic made using VOSviewer). The size and color of the nodes represent the number of references that are shared among the analyzed papers. The strength of a link indicates the number of cited references that the two publications have in common.

VOSviewer is a bibliometric analysis tool used to correlate, construct, summarize, and visualize bibliometric data obtained from different publication databases [30,31]. The advantage of VOSviewer is that it can indicate the interactive relationship between a co-authorship network and bibliographic coupling analysis, which gives an overview to identify potential research hotspots for future research [32]. Here, we used VOSviewer software to correlate the bibliometric data and visualize the interlinks between different piezo scaffolds with their biological properties.

After obtaining and creating the bibliographic library, we performed three bibliographic coupling analyses, as demonstrated in Figure 3C–F. Figure 3C,D demonstrate the bibliometric coupling analysis diagrams, where different piezoelectric biomaterials were explored for several tissue engineering applications and corelated with their reported cellular functionality. For example, most of the piezoelectric biomaterials were fabricated in the three most preferred forms i.e., nanofibers, hydrogels, and 3D-printed scaffolds composed of polymeric materials for the support of tissue repair and regeneration. In general, it was observed that the applications of piezoelectric biomaterials in tissue engineering to a range of hard and soft tissues were closely linked to the study of cellular responses such as adhesion, differentiation, and proliferation.

Most studies explored the cell adhesion and proliferation behavior of piezoelectric-based nanofiber scaffolds, whereas other fabricated forms such as nanoparticles and hydrogels were explored for their superior biocompatibility and stem cell proliferation. Notably, there were limited studies on antimicrobial activity and related mechanisms. Interestingly, studies related to bone and cartilage tissue engineering mainly correlated with the mechanical strength, durability, osteoconduction, osteointegration, osteogenesis, and cell differentiation properties of piezo-materials (Figure 3E), whereas in the case of skin tissue engineering or wound healing applications, it was linked with the cell adhesion, proliferation, migration, and stem cell differentiation characteristics of biomaterial scaffolds due to the piezoelectricity (Figure 3F). The work carried out related to dental tissue engineering was also linked with the stem cell differentiation, mechanical strength, cell adhesion, proliferation, and antibacterial effect-based bioactive properties of piezo-materials. However, it is important to note that this bibliometric analysis was only based on scientific publications in the English language and herein, the sole intention herein is to offer a bird's eye view of the overall research trends on piezoelectric-based biomaterials before delving deeper into the materials science in later sections.

Piezoelectric responses in these studies were characterized using state-of-the-art approaches such as piezo force microscopy (PFM) [5] or PiezoMeter by measuring piezoelectric strain coefficients and piezoelectric charge constants (d), piezoelectric voltage coefficients (g), and electromechanical coupling coefficients (k). Among these, determination of the piezoelectric charge coefficients (d) and piezoelectric voltage coefficients (g) was the most common. It is also important to mention that piezoelectricity is a third rank tensor and thus all the coefficients are dependent on the directions of the applied pressure [27]. Taken together, this bibliometric analysis shows that there is emerging interest in developing piezo-material-based biomaterial scaffolds for multifunctional applications. However, future research should investigate the exact mechanisms underlying the tissue engineering responses of piezoelectric materials.

4. Piezoelectric Biomaterials for Engineering the Dentin-Pulp Complex

Over the years, many materials, including ceramics, polymers, doping elements, and synthetic amino acid/polypeptides, have been discovered to have piezoelectric properties. In this part of the review, we critically discuss piezoelectric materials that have potential to be developed for dental tissue engineering (Table 2).

Table 2. Piezoelectric biomaterials and their potential applications.

Piezoelectric Biomaterials	Scaffolds	Piezoelectric Coefficients (d_{33})	Potential Applications in Dental Tissue Engineering
Barium Titanate	Composites Dental cement Nanoparticles	191 pC/N [33]	Cell adhesion and proliferation Stem cell differentiation Dental cement Medical imaging Antimicrobial Remineralization
Zinc oxide	Dental cement Nanorods Nanoparticles	3.62 pC/N [34]	Antimicrobial agent Drug delivery Stem cell differentiation Dental cement Medical imaging
PLLA	Composites Hydrogel Nanoparticles	−10 pC/N [35]	Cell adhesion and proliferation Stem cell differentiation Angiogenesis Drug delivery
PHBV	Nanofibers Composites Nanoparticles	1.3 pC/N [36]	Cell adhesion and proliferation Drug delivery
Collagen	Nanofibers 3D-bioprinted scaffold Sponge Nanogel Nanoparticles	0.3 pC/N [36]	Cell adhesion and proliferation Stem cell differentiation Angiogenesis Drug delivery Tissue regeneration Pulp-dentin repair/regeneration
Cellulose	Nanofibers 3D-printed scaffold Sponge Nanogel	0.2–0.4 pC/N [37]	Cell adhesion and proliferation Stem cell differentiation Angiogenesis Drug delivery
Chitosan	Nanofibers 3D-printed scaffold Sponge Nanogel Nanoparticles	0.2–2.0 pC/N [38]	Cell adhesion and proliferation Stem cell differentiation Angiogenesis Drug delivery Tissue regeneration Pulp-dentin repair/regeneration
Silk fibroin	Fibers Porous scaffold	d_{14}= −1.5 pC/N [39]	Cell adhesion and proliferation Drug delivery Tissue healing and regeneration

4.1. Piezoelectric Ceramics

One of the earliest piezoelectric materials was lead-zirconium-titanate or PZT, which exhibits an easy dipole orientation and possesses a high piezoelectric coefficient of about 250–350 pC/N due to its ABO_3 perovskites crystal structure (where A and B are cations) [40,41]. However, it is not suitable for biological applications due to the cytotoxic nature of Pb^{2+} ions [41,42] and hence is not discussed in this review. However, it is notable that lead-free piezoceramics such as Li-modified $(Na,K)NbO_3$(LNKN), (Bi, Na)TiO_3 (BNT), $(Na,K)NbO_3$(NKN), and tungsten bronze (TB) have been developed to circumvent the challenge of toxicity, but their biological applications remain to be elucidated.

4.2. Barium Titanate (BT)

The piezoelectric property of barium-based ceramic materials was discovered during the poling of BT. Electrical poling or corona poling is the process of phase transition achieved by the application of a high electrical field (1–30 kV) at temperatures higher than the Curie temperature [19,41]. BT offers a superior advantage for biological use than other piezoceramics such as PZT due to its excellent biocompatibility even at concentrations >100 μg/mL [43]. The piezoelectric coefficient d_{33} of BT is <350 pC/N. The incorporation of BT nanoparticles in a PLGA polymer matrix improved its cell proliferation, osteoconductive, and osteointegration behavior in osteoblasts and osteocytes, favoring bone tissue engineering [43]. The incorporation of BT nanoparticles as a filler material in the traditional dental resin composite [44] showed antibacterial and remineralization effects. BT-incorporated denture polymers, such as polymethyl methacrylate, eradicated fungal biofilms and potently killed Candida albicans [16] due to the piezoelectric charges generated by the barium titanate nanoparticles during simulated masticatory forces, which resulted in the generation of reactive oxygen species (ROS) and consequent upregulation of the superoxide dismutase gene (SOD5) [16].

A piezoelectric $BaTiO_3$-hydroxyapatite-based nanocomposite platform was reported by Dhall et al. [17] to overcome biofilm-associated infections and consequent failures in medical devices (Figure 4). They showed that the addition of piezoelectric $BaTiO_3$ nanoparticle in a hydroxyapatite scaffold resulted in dose-dependent activity against *Streptococcus mutans* biofilms. They observed a 10-fold reduction in colony-forming units (CFU) compared to the pristine scaffold. Furthermore, it was demonstrated that an increase in the concentration of the piezoelectric $BaTiO_3$ nanoparticles from 10 to 30 wt% in the scaffold increased the negatively charged surface energy, creating an unfavorable condition for bacterial adhesion [17]. Nonetheless, Fan et al. reported the angiogenetic capability of BT under the influence of ultrasound waves in a titanium implant coated with BT nanoparticles for large segmental bone defects [19]. Therefore, having multifunctional properties (antibacterial, antibiofilm, anti-inflammatory, mineralization, and angiogenesis), long-width piezoelectric BT-based piezo-platforms have great potential for further research on regeneration of the dentin-pulp complex.

4.3. Zinc Oxide (ZnO)

Despite the well-explored biological applications of ZnO due to its antibacterial property, biocompatibility, and its favorable role in cell adhesion and differentiation, the investigation and application of its piezoelectric properties remain limited. ZnO crystals demonstrate a piezoelectric d_{33} coefficient of 3.62 ± 0.5 pC/N [20] because of their hexagonal asymmetric wurtzite crystal structure and polarized crystal surface [19,21]. A recent study by Bhang et al. [5] demonstrated the effect of piezoelectric ZnO nanorods in wound healing and skin tissue regeneration in both in vitro and in vivo mice models. They developed a multi-layered dermal patch by reinforcing a polydimethylsiloxane (PDMS) matrix with ZnO nanorods through pin coating. Mechanical rubbing with a soft velvet cloth was used to induce the directional alignment of the dipole. By varying the concentration of ZnO nanorods to 54.8% and 95.2%, they obtained a range of piezo potentials from 300 to 900 mV [5]. The fabricated patch significantly increased piezoelectric biomarkers, including PCNA, TGF-β, COL-III, COL-IV, and α-actin. It was also shown to upregulate several marker genes such as CD68, VEGF, and CD99 in an athymic mice model. Taken together, these findings signify that the piezoelectric charges from ZnO are involved in various molecular pathways that regulate cell migration, tissue granulation, cell proliferation, and differentiation, which are essential for skin regeneration.

A study on the size-dependent cytotoxic behavior of ZnO [18,21] reported that macro- and micro-range (>1 μm) ZnO particles do not have any toxicity, but nanoparticulate ZnO (i.e., 20–200 nm) exhibits cytotoxicity above a 0.2 μg/mL concentration in squamous cell carcinoma (HNSCC) and HepG2 cells in vitro due to significant reactive oxygen species (ROS) generation [45,46]. Nevertheless, this drawback could be overcome by surface and chemical modifications. For instance, Ramasamy et al. [46] found that a thick coating of

SiO_2 on the surface of 20 and 50 nm sphere-like ZnO NPs improved its cytocompatibility on human skin dermal fibroblast neonatal (HDFn) cells compared to bare ZnO NPs [46]. Furthermore, the modified hydrophilic surface of the SiO_2 coating stabilized the ZnO NPs due to loosened aggregation. The coated ZnO NPs had a lesser degree of LDH leakage, ROS production, and LPO release compared to pristine ZnO NPs [46]. However, the impact of these surface modifications on the piezo property of nanoparticles remains unknown.

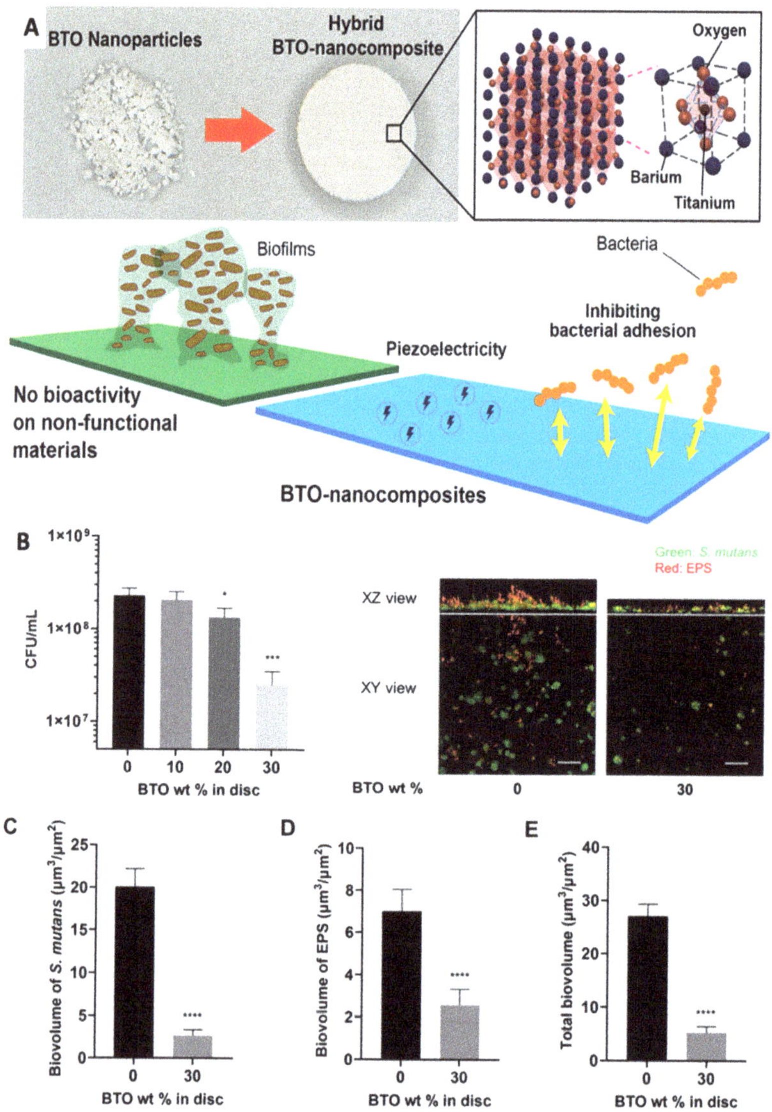

Figure 4. (**A**) Bimodal piezoelectric barium titanate nanocomposite exhibits antibiofilm activity; (**B**) Dose-dependent anti-biofilm activity and confocal images of *Streptococcus mutans* biofilms after 18 h on the BT/hydroxyapatite piezoceramic discs. (**C–E**) Quantified biovolume of *S. mutans*, EPS, and total biovolume (sum of *S. mutans* and EPS) in the biofilm, Statistics: *t*-test with *, *** and **** represents $p < 0.01$, $p < 0.001$ and $p < 0.0001$ respectively; Reprinted with permission from [17], Copyright 2021, [17], American Chemical Society, with minor modifications.

4.4. Piezoelectric Polymers and Composites

Piezoelectric polymers have several advantages over ceramic- and metal-based piezoelectric materials such as facile fabrication, broad tunability of physiochemical properties by surface functionalization, chemical modification, and the ability to fabricate different types of scaffolds such as nanoparticles, nanofibers, nanorods, hydrogels, composites, and cryogels [18,29]. Moreover, their low cytotoxicity and physiological roles such as cell adhesion and proliferation [5], stem cell differentiation [47,48], and ultrasound-mediated mechanotransduction ability [49] endow them with unique advantages in biomedical applications.

4.4.1. Synthetic Polymers

- Polyvinylidene fluoride (PVDF) and its derivatives

PVDF has been used as the 'gold standard' piezoelectric polymer for various tissue engineering applications. Along with its piezoelectric property, PVDF has unique advantages such as antioxidant behavior, high strength, and high thermal, chemical, and hydrolytic stability [3,48]. PVDF exhibits piezoelectric properties due to its polarized molecular structure. The general molecular formula of PVDF is ($-CH_2-CF_2-$) n. Typically, PVDF exists in five polymorphic forms, namely α, β, γ, δ, and ε phases, depending on the nature of the tacticity of the hydrogen and fluorine atoms [50]. Among these, only the β phase shows inherent piezoelectricity due to its all-trans molecular configuration and in this atomic configuration, all the $-CH_2-$ dipoles are perpendicular to the $-CF_2-$ repeat units, which results in an inherent electric dipole moment [50].

Several techniques such as corona poling, uniaxial and biaxial drawing, high voltage electrospinning, and annealing facilitate the phase transformation and increase the β phase concentration in PVDF [16,51]. PVDF shows an average piezoelectric d_{33} coefficient of 20 pC/N [48] due to the presence of highly electronegative F atoms and its trans-gauche-trans-gauche' (TGTG') atomic configuration in a centrosymmetric unit cell [51]. The negatively charged surface of PVDF scaffolds is induced through the corona poling process to promote better cell adhesion and proliferation in C2C12 mouse myoblast cells compared to non-poled PVDF specimens for skeletal muscle regeneration [52]. More recently, it was demonstrated that poled β-PVDF samples have better protein adsorption and osteogenic differentiation on human adipose stem cells (hASCs) compared to unpoled scaffolds [53]. It was also reported that PVDF fibrous membrane as a scaffold for growing and recapitulating the multi-layered chondrocytes resulted in significantly greater gene expression for fibronectin and integrin α-10 in chondrocytes that adhered to the PVDF surface [53]. Furthermore, this study demonstrated that chondrocyte cell sheets had a similar phenotype to the regular ones and had increased gene expression of SOX9 and Col XXVII [53]. In a different study, the mineralization potential of PVDF scaffolds was revealed, with the amount of formed mineral proportional to the magnitude of external mechanical stimulation [54].

Two-dimensional and three-dimensional PVDF nanofiber scaffolds show differential effects on human-induced pluripotent stem cells (iPSCs) [34]. iPSCs seeded on 3D fibrous scaffolds showed significantly greater osteogenic-related genes and protein expression compared to the 2D scaffold, resulting in superior osteoinductive effects and better bone differentiation. Another remarkable study by the same group reported the potential of PVDF-polyaniline (PANI) piezoelectric electrospun scaffold for the differentiation of stem cells derived from dental pulp (DPSCs) [55]. To the best of our knowledge, this is the only study to date that has indicated the effect of piezoelectricity on the osteogenic differentiation of dental stem cells. This study applied a low-frequency pulsed electromagnetic field (PEMF) as a source of mechanical stimulation for the piezo-scaffolds and observed that DPSCs seeded on the scaffold exhibited better cell adhesion, increased alkaline phosphatase activity, a higher calcium content, and significantly higher osteogenic gene expression compared to unexposed samples (Figure 5) [56].

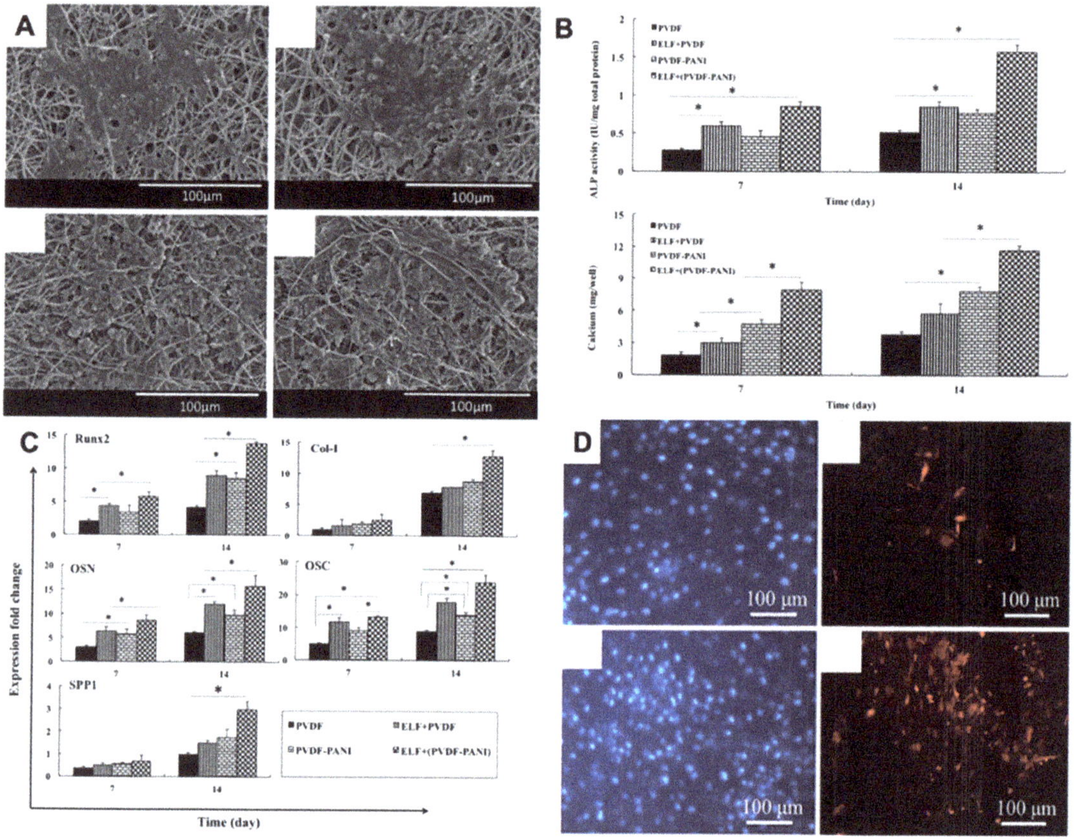

Figure 5. (**A**) SEM images of DPSC-seeded electrospun PVDF scaffold after PEMF exposure; (**B**) Alkaline phosphatase and calcium content assays of the differentiated DPSCs under osteogenic medium; (**C**) Different relative gene expression of DPSC-seeded electrospun PVDF scaffold at 7 and 14 days in the absence and presence of PEMF, the significant differences ($p < 0.05$) between groups are indicated with * sign; (**D**) Immunocytochemistry (ICC) staining for osteocalcin protein in the differentiated DPSCs. Reprinted with permission from [56]. Copyright 2019 [56], Taylor & Francis Online, with minor modifications.

Chemical derivatization of pristine PVDF polymer such as P(VDF-TrFE), a copolymer of vinylidene fluoride (VDF) and trifluoroethylene (TrFE) [57,58], results in an enhanced electromechanical conversion compared to PVDF [59]. For instance, Hunage et al. reported that biaxially oriented poly (vinylidene fluoride) exhibits a higher piezoelectric coefficient ($d_{33} = -62$ pC/N) than pristine PVDF [60]. Electrospun nanofibrous mats of P(VDF-TrFE) were shown to regenerate neuronal tissue from PC-12 cells upon ultrasound stimulation [4]. Notably, the neurite outgrowth was uniform in all directions on the piezo scaffolds compared to the neural growth factor added media. It has also been demonstrated that human mesenchymal stem cells (MSCs) cultured on a thermally poled piezoelectric PVDF-TrFE nano-fibrous scaffold exhibited tissue-specific chondrogenesis and osteogenesis as confirmed by the GAG content, collagen type II to I ratio, ALP-mediated mineralization, and osteogenic gene expression [61].

Dental pulp stem cells (DPSCs) grown on PVDF-polyaniline piezoelectric nanocomposite scaffolds were shown to have improved osteoinductive capability. This study demonstrated a higher level of DPSC adhesion, alkaline phosphatase activity, calcium content, and osteogenic gene expression under a low frequency pulsed electromagnetic field (PEMF) [56]. Therefore, blending a conductive polymer such as polyaniline with a piezoelectric polymer could be a useful approach for improving cell attachment and stem cell differentiation while it remains unknown whether the cells can be differentiated to preferred phenotypes by modifying the piezoelectric behavior.

Multifunctional piezoelectric-based smart dental implants (SDIs) composed of PVDF and $BaTiO_3$ have been shown to exhibit anti-inflammatory activity in addition to its concomitant use with photo-biomodulation (PBM) therapy by energy generated from normal human oral motions such as chewing and brushing (Figure 6). SDI was shown to exhibit higher cell viability and anti-inflammatory activity upon harvesting the electrical charge that accumulated on the piezoelectric scaffold [62]. This work demonstrated that the piezo scaffold was able to generate light in situ for photo-biomodulation (PBM) therapy, which regenerates and restores the damaged peri-implant soft tissue. Traditional NIR-based photodynamic therapy (PDT) for peri-implant soft tissue regeneration requires 0.8 V electrical stimulation, which can be generated from the piezoelectric scaffold from 60 N mechanical pressure of chewing motion or 90 N of brushing motion [62]. Thus, this groundbreaking study set the stage for integrating multifunctional and smart therapeutics using piezoelectric materials in oral healthcare.

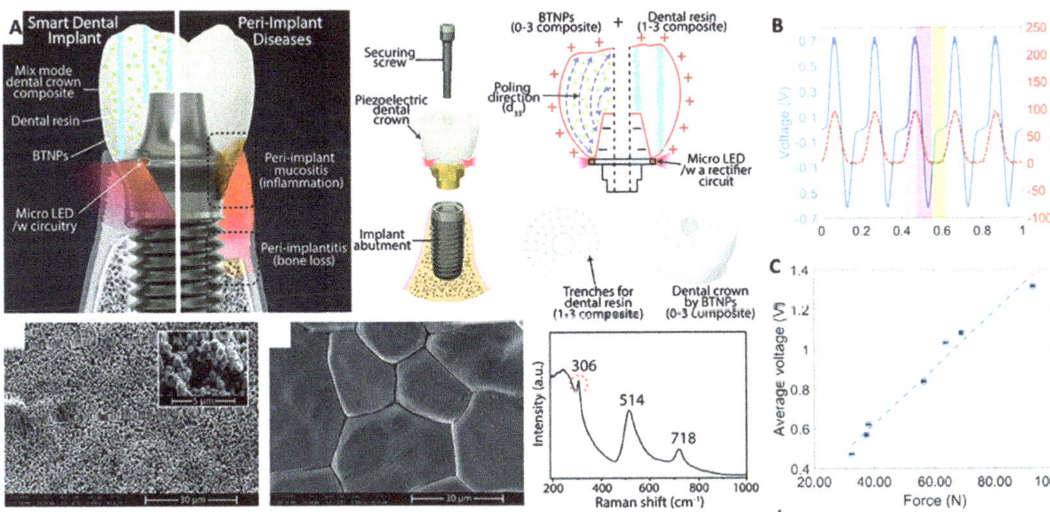

Figure 6. (**A**) Schematic illustration and SEM images of Piezoelectric Smart Dental implants (SDIs) capable of generating light under physiological chewing and brushing motion for photo-biomodulation therapy; (**B**,**C**) Electrical voltage generated from the piezoelectric-SDI under chewing motion (the applied force was ≈90 N at a frequency of 5 Hz) and average voltage outputs of SDI under soft food chewing motions that ranges from 30 to 100 N (f = 5 Hz). Reprinted with permission from [62], Copyright 2022, [62], Wiley Online library, with minor modifications.

Moreover, all the above findings indicate that PVDF serves as a center of attention of piezoelectric materials due to its ease of processability, stable piezoelectric response, and good biological characteristics. However, limited attention has focused on tuning the chemical properties of PVDF instead of blending with other materials; for instance, surface functionalization or conjugation of biomolecules could be a better approach to explore the multifunctionality of PVDF-based piezo scaffolds. We believe that the next-generation bio-piezoelectric platforms can be developed by strategically combing the physical piezoelectricity with the chemical properties of piezoelectric biomaterials.

- Poly-L-Lactic Acid (PLLA)

PLLA is one of the best-known biodegradable polymeric biomaterials used for dental tissue engineering [63,64]. There are four polymorphic forms of PLLA—α, β, δ, and γ phase, all of which exhibit piezoelectricity [65]. The helical topography and quasi-crystalline nature are responsible for the piezoelectric property of PLLA, resulting in a piezoelectric shear coefficient (d_{14}) of −10 pC/N [65,66]. Electrospun nanofibers of PLLA have been shown to induce DPSCs' differentiation into mature odontoblasts (Figure 7A,B) and recapitulate the dentin-pulp histoarchitecture in vitro [48]. However, whether the piezoelectric properties of PLLA contribute to such effects remains unknown.

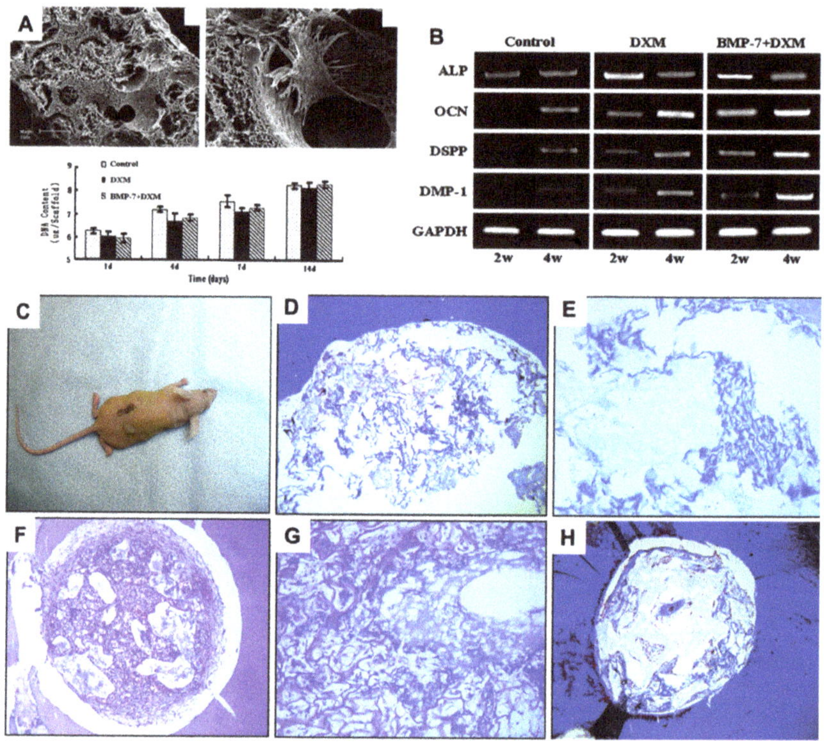

Figure 7. (**A**) SEM images of PLLA-nanofibrous scaffolds loaded with DPSCs cultured in vitro for 3 days and proliferation characteristics of DPSCs; (**B**) Gene expression of DPSCs grown on PLLA scaffolds for 2 and 4 weeks. Reprinted with permission. Copyright 2010 [41] Elsevier, with minor modifications. (**C**–**H**) Histological images of collagen scaffold, DPSCs, and DMP1 scaffold constructs implanted in vivo for 8 weeks. Reprinted with permission from [35]. Copyright 2018 [35], Journal of Endodontics, with minor modifications.

An advantage of the PLLA-based piezoelectric scaffolds is that they do not require additional electrical or mechanical poling to enable piezoelectricity. Comparing the proliferation and maturation of DPSCs on various scaffolds, Chandrahasa et al. showed that this nonpolling requirement capability is influenced by the chemical composition of the scaffolds and that PLLA scaffolds showed greater mineralization ability than bovine collagen and calcium phosphate scaffolds [67]. A recent study by Das et al. [68] showed the osteogenic differentiation ability of piezoelectric PLLA nano-fibrous scaffolds from adipose stem cells (ADSCs) and bone marrow stem cells (BMSCs) under ultrasound stimulation. They reported that a piezoelectric PLLA scaffold with ultrasound stimulation exhibited greater collagen 3.6 gene expression in a bone defect mice model than without ultrasound.

It is noteworthy to state that the upregulation of the collagen 3.6 gene, which is normally present in mature osteoblasts, directly implies greater osteoblastic activity. Moreover, PLLA-based scaffolds show controlled biodegradation and excellent biocompatibility. However, it is noteworthy that the biodegradation of PLLA forms lactic acid, which may have a negative impact on the mineralization of dental hard tissue [69].

- Poly-3-Hydroxybutyrate-3-Hydroxyvalerate (PHBV)

PHBV has received considerable attention due to its accepted biocompatibility, biodegradability, and appropriate mechanical properties and thermoplasticity for biomedical applications [70]. Interestingly, PHBV also possesses a piezoelectric d_{33} coefficient of 1.3 pC/N, which is almost equivalent to type 1 collagen in human bone [71]. Due to these characteristics, PHBV-based piezoelectric scaffolds have been investigated for bone tissue engineering. Kose et al. reported the cartilage tissue regeneration ability of collagen-PHBV scaffolds [71]. A study by Jacob et al. [72] showed that the nano-fibrous scaffolds of BT-nanoparticle-reinforced PHBV composites exhibit higher chondrocytes and Col-II gene expression activity. They also pointed out that the reinforcement of BT-NPs as a filler increases the strength of the pristine PHBV scaffold up to 20%. This study indicated that it is possible to regenerate or repair any tissue without using any type of chemical or biological factors such as growth or transcription factors by solely utilizing piezoelectric scaffolds.

4.4.2. Natural Biopolymers

- Collagen

Being the most abundant natural biopolymer, collagen is considered one of the best biomaterials for tissue engineering applications, including dental tissues [73,74]. It is known that collagen type I in the dental matrix serves as the base material for the calcification of dental tissue [75,76]. The piezoelectric shear coefficient (d_{15}) of collagen type I fibril was found to be 0.51 pm/V, as quantified by PFM [77].

There is evidence that demonstrates that collagen scaffold induces the regeneration of odontoblasts and encourages the formation of odontoblasts and adhesion with the pulp [78]. Over the past few years, numerous studies have reported a multitude of applications of collagen scaffolds, collagen gels, and sponge in dentin-pulp complex regeneration, stem cell differentiation, and cellular proliferation [35]. One such study [35] demonstrated that after 6 weeks of seeding DPSCs on a collagen substrate, a physiological-mimicking matrix architecture was formed (Figure 7C–H). In addition, a DPSC-seeded 3D collagen scaffold can proliferate and differentiate into odontoblasts [79]. However, they did not decipher the role of piezoelectricity in this differentiation. Piezoelectric collagen-hydroxyapatite composites not only have sufficient mechanical strength, biocompatibility, and low antigenicity but also exhibit better cell adhesion, proliferation, and bone healing properties [80]. As a functional piezoelectric biomaterial, collagen-based piezo-scaffolds can provide the merits of good biocompatibility and degradation, but it also true that most collagen-based scaffolds have the burden of a stable piezoelectric response and the direction dependability of piezo-response of the collagen fibers also has a great impact on the clinical translation. Future research should be focused on the tunability of collagen-based piezo-scaffolds as these aspects can make it a gold standard piezoelectric material from biological origins.

- Chitosan

Chitosan is an abundant biopolymer, generally obtained from the deacetylation of chitin, which is found in the exoskeleton of insects and mollusks [81]. It is a remarkably well-explored biomaterial in various applications, including drug delivery, tissue engineering, wound healing, and biosensors [36,82]. Nevertheless, to facilitate an improvement in the mechanical and biological properties, chitosan is often blended with other materials. For instance, chitosan-carboxymethylcellulose (CMC) hybrid scaffolds demonstrated enhanced proliferation and significantly higher gene expression of osteonectin and dental sialophosphoprotein compared to a native chitosan scaffold [83]. Yang et al. reported distinct dental stem cell adhesion, proliferation, and differentiation properties of a BMP-

7-loaded chitosan-collagen scaffold [74,84]. In addition, Liao et al. developed a bioactive chitosan scaffold with β-tricalcium phosphate (TCP), which promoted the vascularization of human periodontal ligament cells (HPLCs) in vivo [85].

Chitosan nanoparticles have also been added to endodontonic sealers for antibacterial purposes. For example, the incorporation of 2% wt/vol. chitosan nanoparticles into three commercial endodontic sealers (AH Plus-Dentsply DeTrey, Konstanz, Germany), Apexit Plus-Ivoclar Vivadent, Schaan, Liechtenstein, and MTA Fillapex-Angelus, Londrina, Brazil) resulted in superior antifungal activity than the sealers used alone [86]. Detailed investigations of chitosan nanoparticles by Kishen's group [87] demonstrated their antibacterial and antibiofilm efficacy using dentin infection models. It was shown that polycationic chitosan nanoparticles interact with the negatively charged bacterial cell surface to cause bacterial killing and eradication of biofilms [87]. Excitingly, chitosan was recently shown to possess piezoelectric characteristics due to its ortho-rhombic crystal structure with the $P2_12_12_1$ space group and non-centrosymmetric attributed to the glucosamine monomer [88]. Chitosan exhibits a range of piezoelectric d_{33} coefficients from 0.2 to 1.5 pC/N, depending on its source and degree of deacetylation [85,89]. Despite such promising findings in the above works, a cause–effect relationship between tissue regeneration and piezoelectric behaviors has not been reported for chitosan.

- Cellulose

Cellulose is the most abundant natural polymer, which exhibits excellent sustainable properties such as biodegradability, eco-friendly, low-cost production, excellent biocompatibility, and outstanding mechanical properties [90]. For these reasons, cellulose was explored as a center of attention to decipher its piezoelectric capability. Generally, cellulose is composed of a linear chain of glucose molecules with three side hydroxyl groups (-OH), with the glucose moieties connected by a β-1,4-glycosidic linkage [91]. The strong H-bonds between the side -OH groups give a unique preposition to the cellulose moieties to form a highly ordered crystalline structure. Usually, cellulose crystals exist in four different forms, i.e., cellulose I, II, III, and IV [92,93]. Among them, the most common naturally originating cellulose I is present in two different polymorphs forms: triclinic type I_α and monocline type I_β, depending on the source of the extraction [94,95]. Interestingly, another type of crystal, cellulose II, which is monoclinic in nature, is converted into cellulose I by dissolution and alkali treatment [90]. The piezoelectricity of cellulose results from the net dipole moment of polar -OH groups present in the triclinic type I_α polymorph, which is arranged in a non-centrosymmetric order [95]. The piezoelectric property of cellulose was first explored by Fukada [96]. It was reported that the longitudinal piezoelectric d_{33} coefficient of natural cellulose is about 0.4 pC/N [37].

Recently, cellulose, which is extracted from bacterial species, has been reported to possess excellent mechanical properties, high water holding capability, and outstanding suspension stability features [97]. For instance, An et al. fabricated a bacterial cellulose membrane for the guided bone regeneration (GBR) using electron beam irradiation techniques [98]. They demonstrated that electron irradiation of a bacterial cellulose membrane resulted in enhanced in vitro cell viability, adhesion, and proliferation in NIH3T3 cells and in vivo bone regeneration on rat calvarial defect models compared to non-irradiated samples [98]. Another study [99] reported that reinforcement of micrometric particles of cellulose into silicate dental cement resulted in a shorter setting time, enhanced compressive strength, and enhanced cell adhesion and proliferation [99]. However, cellulose remains to be exploited as a piezoelectric material for the regeneration of dental tissues.

- Silk Fibroin

Being a natural biopolymer with good biocompatibility, excellent mechanical strength, and controlled biodegradation properties, silk has been explored as a multifunctional biomaterial for different biomedical applications, including drug delivery, tissue engineering, and regenerative medicine. For example, Woloszyk et al. [100] reported the neovascularization ability of DPSCs and gingival fibroblasts on a silk fibroin-based scaffold. They demonstrated that both cells have equal affinity to the formation of attracting blood vessels towards the damaged tissue microenvironment [100]. Jiang and coworkers [38] also investigated the proliferation and differentiation properties of DPSCs over a 3D-printed collagen/silk fibroin scaffold. They reported that after 1–5 days of incubation, it was observed that DPSCs seeded on a scaffold exhibited better cell adhesion and enhanced ALP activity, which induced DPSCs differentiation [38]. Kweon et al. [101] studied the effect of the addition of silk fibroin and hydroxyapatite coating on dental implants. They observed that after 6 weeks of implantation in a rabbit tibia model, the combined silk fibroin and hydroxyapatite coating groups had more new bone formation and bone-to-implant contact compared to uncoated and only hydroxyapatite-coated implants [101]. Nevertheless, a biomimetics approach was investigated by Huang et al. [102] for the creation of biominerals using a combination of spider silk and dentin matrix protein 1 (DMP-1). To achieve this, a novel spider-like domain and a domain of DMP-1 were cloned and expressed, and the two domains were then used for self-assembly and nucleation of hydroxyapatite [102].

Natural silk fibroin is a special type of block copolymer composed of two different heavy (~370 kDa) and light (~26 kDa) chains linked by disulfide bonds [103]. The heavy chain consists of alternating hydrophobic, repetitive oligopeptides that are separated by smaller charged and amorphous sequences. The hydrophobic domain is rich in alanine and glycine amino acids while the hydrophilic spacers give the heavy chain a polyelectrolyte nature [104]. Naturally, silk fibers are available in two different polymorphic forms, i.e., silk I and silk II [105]. Among them, silk II is present as a pleated, antiparallel β-sheet secondary structure with a monoclinic unit cell. The piezoelectric potential of the silk fibroin fibers mainly originates from its β-sheet content of silk II polymorphs. The uniaxially oriented polycrystalline silk fibers exhibit shear piezoelectricity, which indicates that the silk fibers generate electricity upon the exposure of certain shear stress perpendicular to its orientations [104,105]. A recent study by Yucel and colleagues [39] reported evidence of the structural origin of the piezoelectricity of silk fibroin. They evidenced that silk fibers exhibit a shear piezoelectric coefficient $d_{14} = -1.5$ pC/N after processing using the zone drawing method [39]. They also reported a correlation of the β-sheet content with an increasing draw ratio and the simultaneously increasing degree of the orientation of β-sheet crystals [39]. Thus, with the synergy of regeneration, mineralization capability, and piezoelectric potential generation, silk fibroin could serve as a potential piezo-biomaterial for dentin-pulp complex regeneration.

4.5. Amino Acids, Polypeptides, and Proteins

Intrinsic polar groups such as amino (-NH_2) and carboxylic acid (-COOH) in the molecular structure of natural amino acids endow them with a unique advantage to exhibit piezoelectricity [39]. Most amino acids have been shown to possess piezoelectric potential due to their distinct crystal habits such as a right-handed (D) or left-handed (L) form [106]. Except for glycine, all amino acids have a chiral center, which results in the formation of non-centrosymmetry in the crystal lattice of the amino acids. This non-centrosymmetry of the groups results in the generation of piezoelectric tensors. However, most of the L or D-form amino acids exhibit shear piezoelectric tensors such as d_{14}, d_{25}, and d_{36} rather than longitudinal piezoelectric tensors such as d_{11}, d_{22}, or d_{33} [106]. It is important to note that the possession of a longitudinal piezoelectric coefficient is essential for real-world biological applications since real-life mechanical forces are perceived either in the compressive or tensile direction. Thus, attempts have been made to fabricate unidirectional and longitudinal piezoelectric amino acid crystals.

The extracellular matrix of mammalian tissues also comprises electroactive nanocrystalline polypeptide or protein molecules such as keratin, collagen, elastin, and glycosaminoglycans [107,108]. Piezoelectric phenomena in keratin were first reported by Martin when he observed the generation of a static electric potential from a bundle of wool (which is primarily composed of keratin) compressed within two brass plates. Keratin exhibits piezoelectric characteristics due to its highly ordered α-helical structure and the dipole originates from the hydrogen bond presence between the amine (-NH_2) and carbonyl (-C=O) groups [7].

Recently, Guerin et al. studied the presence of a longitudinal piezo-response in β and γ-glycine due to their orthorhombic crystal orientation and validated the piezoelectric coefficient (d_{16}) in β-glycine of about 2×10^2 pm·V^{-1} [109]. They also reported a drop-casting solution-based facile fabrication method for the fabrication of amino acid crystal films with longitudinal piezoelectricity [109]. Density functional theory (DFT) based computational calculation was used to quantify the theoretical value of piezoelectric coefficients for several orthorhombic L-amino acid crystals (Table 3) such as threonine, asparagine, glutamine, histidine, proline, methionine, and isoleucine [110–112].

Table 3. Reported piezoelectric coefficient value of amino acids/peptide-based piezoelectric biomaterials, measured using PFM or DFT-calculation. The subscript in the brackets indicate the respective directions.

Amino Acid	Piezoelectric Coefficient (pC/N)	Reference
Threonine	4.9 pC/N (d_{36})	[111]
Proline	27.75 pC/N (d_{25})	[111]
Asparagine	13 pC/N (d_{16})	[111]
Histidine	18 pC/N (d_{16})	[111]
Leucine	12.5 pC/N (d_{16})	[112]
Isoleucine	25 pC/N (d_{34})	[111]
Cysteine	11.4 pC/N (d_{22})	[111]
Glycine	178 pC/N (d_{16})	[113]
Alanine	17.75 pC/N (d_{24})	[113]
Poly-Amino Acid/Peptide	**Piezoelectric Coefficient (pC/N)**	**Reference**
Keratin	1.8 pC/N (d_{14})	[106]
Lysozyme	6.5 pC/N (d_{33})	[106]
Diphenylalanine (FF)	80 pC/N (d_{15})	[109]
poly(γ-benzyl-α,L-glutamate) PBLG	25 pC/N (d_{33})	[114]
poly-γ-methyl-l-glutamate (PMLG)	2 pC/N (d_{14})	[115]

5. Fabrication Methods/Delivery Strategies for Piezoelectric Materials

5.1. Electrospun Fibers

Electrospinning is one of the most explored fabrication techniques for the development of nanofibers of piezoelectric materials (Figure 8). Nanofibers are popular scaffolds owing to its high surface area, tunable fiber morphology, ability to tailor the scaffold shape and fiber orientation, ease of surface functionalization, and porous structure, which mimics the natural extracellular matrix [116]. The underlying principle of electrospinning relies on the electrostatic interaction between a concentrated polymer solution and an oppositely charged collector system within a high voltage (1–30 kV) electrical field. Due to the high electric field, the surface tension of the polymer solution decreases and results in surface charge generation followed by Taylor cone formation, which leads to the stretching of the concentrated polymer solution and the formation of nanofiber deposited onto the collector system [117]. The ability to tailor the fiber diameter and morphology depends on several key factors such as the polymer concentration/viscosity, applied voltage, tip to collector distance, flow rate, needle diameter, temperature, and humidity. Different types of collector systems are also used to obtain different alignments and specific geometry of nanofibers; for instance, a commonly used rotating collector is used to align nanofibers in a parallel manner depending on its rotating speed [115,116].

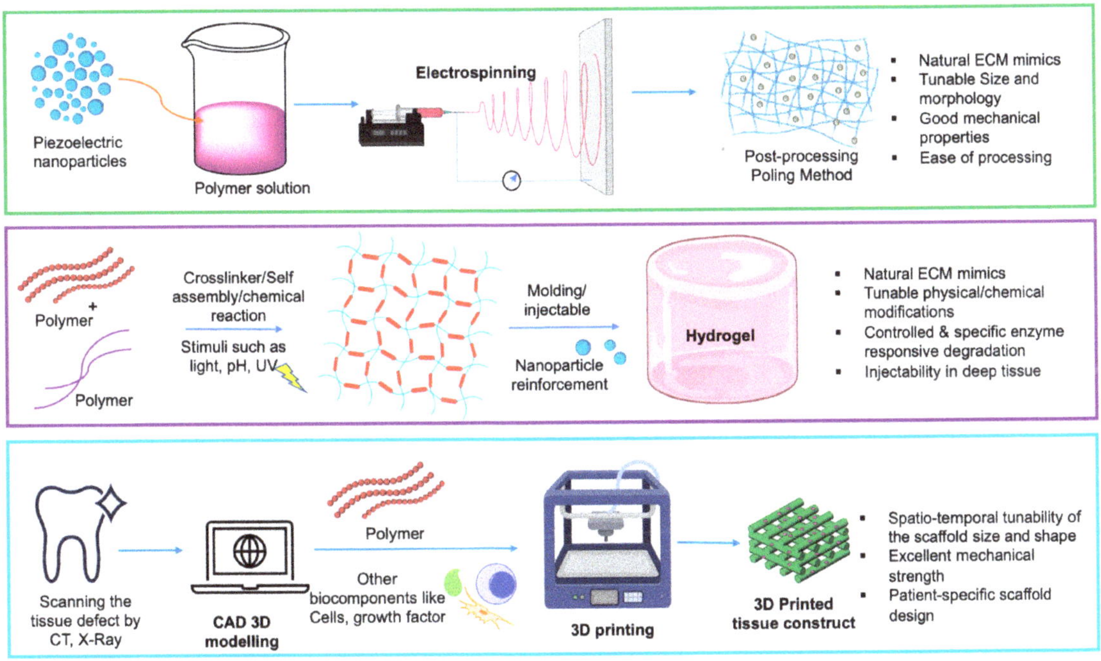

Figure 8. Different methods for the fabrication of piezoelectric biomaterial scaffolds.

Over the years, a plethora of works have reported the use of electrospinning scaffolds in dental applications, including pulp-dentin complex regeneration [118], repair of defects in periodontal tissues such as alveolar bone and periodontal ligament (PDL) [119], and guided tissue regeneration (GTR) membranes [120]. Nevertheless, it is also well evidenced that the reinforcement of nanoparticles with electrospinning scaffolds leads to multifunctionality and increased mechanical properties and tunable biodegradability. One such study reported by Bae et al. demonstrated that an electrospun scaffold of collagen with nano-bioactive glass (nBG) exhibited enhanced cell adhesion and proliferation, better mineralization, and increased levels of odontoblastic gene expression, including DSPP, DMP-1, ALP, OPN, and OCN, which leads to odontogenic differentiation of hDPSCs for dental-pulp tissue regeneration [116]. In another work, Guo et al. fabricated a polyurethane/polyvinylidene fluoride (PU/PVDF) electrospun scaffold for wound healing [121]. They reported that during the electrospinning process, the high electrical field induces a piezoelectric β-crystalline phase of PVDF [121]. Nevertheless, sometimes additional post-processing modification such as corona poling is also carried out to improve the overall piezoelectric performance and increase the d_{33} coefficient. For instance, Das et al. [68] improved the piezoelectric coefficient through the thermal poling process by annealing piezoelectric nanofibers at 105 °C for 10 h. Moreover, the process of electrospinning followed by thermal/electrical poling is a convenient technique to fabricate nanofibers of various piezoelectric polymer materials with diverse applications.

5.2. Hydrogels

A hydrogel is a three-dimensional network of a hydrophilic polymer system with more than a 10% water content in its structure, which in turn shows good biocompatibility due to the high moisture content [122] (Figure 8). Owing to its 3D matrix configuration, it not only mimics the natural ECM microenvironment but also supports cell attachment, proliferation, differentiation, regulating cell behavior, and intracellular signaling, simulating the recovery of the microenvironment of cell life properties [123,124]. A recent review by Ye et al. [124]

reported the versatility of hydrogel scaffolds for several dental purposes such as dental pulp regeneration, periodontal tissue regeneration, and drug delivery. Another recent study by Siddiqui et al. [122] demonstrated the potential of self-assembled peptide hydrogel scaffolds for dental pulp tissue regeneration after pulpectomy, which is a major clinical challenge in endodontics.

Li et al. [123] fabricated a piezoelectric hydrogel of polyacrylonitrile-acrylamide-styrene sulfate-poly (vinylidene fluoride) (PAAN-PVDF) and reported that the fabricated PAAN-PVDF (15%) hydrogel was able to produce an output voltage close to 50 mV when the force was greater than 35 N [123]. However, it is important to note that in the case of the hydrogel matrix system, the output electrical signal resulted from two different aspects: one is the dipole moment of the piezo-material and the shape variable generated from the deformed hydrogel under certain forces. PAAN-PVDF also exhibits large stretchability (~380%) and skin-like ductility and promotes angiogenesis in HUVEC cells under piezoelectric stimulation [123]. Another piezoelectric hydrogel was created by Zhou et al. [123] by combining poly (2-hydroxyethyl methacrylate) (PHEMA) doped with conductive nanoparticles, i.e., graphene oxide (GO) and single-walled carbon nanotubes (SWCNTs). They reported that the piezoelectricity of the resulting hydrogel is directly proportional to the amount of GO reinforcement but has a negative impact in the case of SWCNTs due to the internal slipping within a bunch of the nanotubes [123].

5.3. Additive Manufacturing: '3D Printing'

The exploration of additive manufacturing or 3D printing has been revolutionizing the worlds of tissue engineering and regenerative medicine (Figure 8). Three-dimensional printing gives us unique advantages over other conventional fabrication techniques thanks to its precise patient-specific complex anatomical replica fabrication with minimal efforts [125,126]. In dental applications, the usability of additive manufacturing technology has been explored, from orthodontics, periodontics, and restorative dentistry to endodontics and implant dentistry. For dental applications, different types of 3D printing platforms have been explored for different purposes such as stereolithography (SLA) or fused deposition modeling (FDM), which is mainly used for the fabrication of solid scaffolds, photopolymer jetting, and digital light processing using light cure resins, which have mainly been investigated for tissue engineering and drug delivery purposes [127,128]. The advancement of this growing technology also gives the opportunity to combine spatio-temporal designed scaffolds with cells, growth factors, and biomaterials to fabricate multiscale scaffolds that maximally imitate natural tissue characteristics. This process is also widely known as 3D bioprinting [128]. Additive manufacturing also enables the integration of two different material systems in the same scaffold, which ensures better physiological tissue mimicry and enhances the integration of soft and hard tissue, reducing the chances of stress shielding and improving the rate of tissue regeneration [129].

The application of 3D printing with piezoelectric materials was first reported by Kim's group, who combined BT nanoparticles with PVDF and printed the blend using an FDM system to fabricate a piezoelectric dental implant [17]. Subsequently, after the printing, they also modified the implants using a 1 nm trench laser to fabricate a honeycomb-inspired configuration, which enhances the overall mechanical properties of the implants [15]. Studies have attempted to develop advanced additive manufacturing techniques such as electric poling-assisted additive manufacturing (EPAM) and integrated 3D printing and corona poling (IPC) by combining the corona poling process during 3D printing [129,130]. Such advanced additive manufacturing processes pave the way for rapid scalability of piezo scaffolds by avoiding the post-processing poling stage requirement to enable piezoelectricity [131]. Not only that, after the 3D printing process, one can control the generated electric potential by modulating the amount of mechanical stress, which is also referred to as 4D printing [132]. Collectively, by taking advantage of the 3D printing approach and piezoelectricity, it is possible to fabricate multiscale, multifunctional tissue regeneration strategies.

5.4. Other Methods

Apart from the conventional fabrication techniques, piezoelectric materials are also investigated with other formulation systems. For instance, solvent casting, PDMS molding, and spin coating are other common techniques used to fabricate piezoelectric material-based scaffolds [132,133]. For instance, the piezoelectric dermal patch by Bhang et al. [5] was fabricated using layer-by-layer spin coating followed by PDMS curing. Nevertheless, they also reported that a ZnO nanorod-reinforced PDMS patch was also able to generate piezoelectricity during the normal hand rubbing process, which induces the dipole alignment.

6. Summary, Conclusions, and Future Perspectives

The highly tunable physicochemical properties and multifunctionality of piezoelectric materials are strong advantages when considering this class of material as scaffolds for dental tissue engineering. Despite the evidence that dental tissues such as dentin have inherent piezoelectric properties and the emerging evidence indicating the advantages of piezoelectric biomaterials, research and development in piezoelectricity-based dental tissue regeneration is extremely limited. Even the studies that reported promising activity for biomaterials did not provide experimental proof to establish and apply piezoelectric behavior for this application. Regarding this aspect, Figure 9 provides a bird's eye view of the potential applications of smart piezoelectric materials in dentistry. For instance, piezoelectric materials could be fabricated as nanoparticles or nanorods to form antimicrobial or mineralizing agents or, on the other hand, forms such as nanofibers or 3D-printed scaffolds can be explored for guided tissue regeneration purposes. Nevertheless, combination with small therapeutic agents, peptides, or DNA piezo scaffolds could also be investigated as a next-generation drug delivery platform.

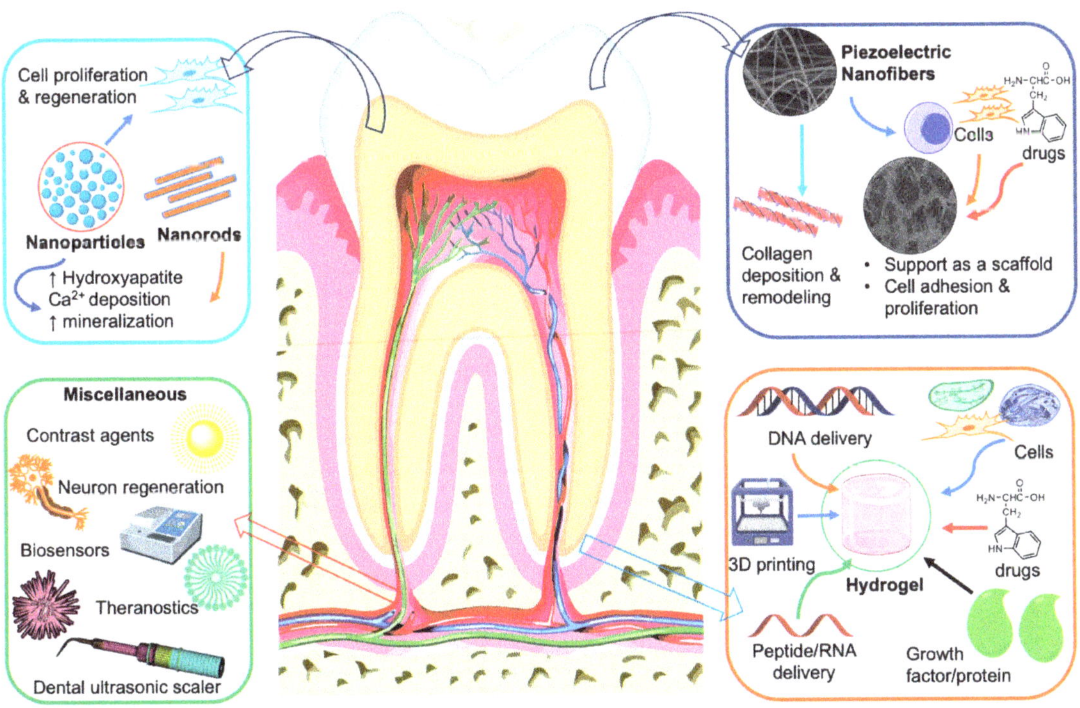

Figure 9. Potential multi-faceted applications of piezoelectric materials in dental tissue engineering.

However, one major challenge in the clinical translation of piezoelectric biomaterial-based scaffold systems is the validation of whether the stress/mechanical load produced during physiological movement is sufficient to generate piezoelectric potential for the specific biological action. For instance, it has been evidenced that a higher electric potential is preferable for mammalian cell adhesion, proliferation, or tissue regeneration and mineralization purposes whereas a lower electric potential is preferable for bacterial killing. For instance, the best antimicrobial effect was shown by 10% BT-reinforced dental resin composite with an electric potential of 1.2 pC/cm^2, which contrasts with the maximum mineralization efficiency exhibited by 60% BT-composites with a higher electric potential, i.e., 3.2 pC/cm^2 [44]. An important research gap in this area is the characterization of whether the same electric potential can kill Gram-positive bacteria, Gram-negative bacteria, and yeasts and eradicate or inhibit biofilms. Another research gap is the lack of a clear elucidation of the mechanisms and molecular signaling pathways by which these materials elicit biological effects both in microbiota and human cells. The long-term biocompatibility of most piezoelectric biomaterials remains to be thoroughly investigated. Furthermore, to develop smart biocompatible piezoelectric scaffold systems, tuning of the piezoelectric characteristics of pristine piezoelectric materials through chemical modification or biomolecule conjugation is an important avenue for future research.

7. Conclusions

The repair and regeneration of the dentin-pulp complex remain a formidable challenge given the sheer biological complexity of this system. The discovery of piezoelectric materials may be considered an important landmark owing to its multifaceted functionalities and excellent tunability. In this review, we discussed how piezoelectric responses generated from piezoelectric biomaterials have been exploited for tissue engineering, antimicrobial, and anti-inflammatory functions. Based on the current review, while it appears plausible that piezoelectric materials may position themselves comfortably as multi-functional biomaterials, the lack of evidence with several outcome measures is also apparent. Despite such voids in the evidence base, these biomaterials may open new doors for successful next-generation regenerative strategies in a field where there is a dire clinical need for innovation and application.

Author Contributions: Conceptualization and Design: S.G., S.O. and P.N.; Literature review: S.G.; Funding acquisition: P.N.; Writing—original draft: S.G., W.Q., Z.Y., S.O. and P.N.; Writing—reviewing, editing and final draft: W.Q., Z.Y., S.O. and P.N. All authors have read and agreed to the published version of the manuscript.

Funding: The authors acknowledge the University of Hong Kong Seed Fund for Translational Research to P.N.

Institutional Review Board Statement: Not applicable.

Informed Consent Statement: Not applicable.

Data Availability Statement: No original datasets were generated in this review.

Conflicts of Interest: The authors explicitly confirm that there are no conflict of interest associated with this publication and there has been no significant financial support for this work that could have influenced its outcome.

References

1. Levin, L.; Day, P.F.; Hicks, L.; O'Connell, A.; Fouad, A.F.; Bourguignon, C.; Abbott, P.V. International Association of Dental Traumatology guidelines for the management of traumatic dental injuries: General introduction. *Dent. Traumatol.* **2020**, *36*, 309–313. [CrossRef] [PubMed]
2. Lawley, G.R.; Schindler, W.G.; Walker, W.A., III; Kolodrubetz, D. Evaluation of ultrasonically placed MTA and fracture resistance with intracanal composite resin in a model of apexification. *J. Endod.* **2004**, *30*, 167–172. [CrossRef] [PubMed]
3. Rajabi, A.H.; Jaffe, M.; Arinzeh, T.L. Piezoelectric materials for tissue regeneration: A review. *Acta Biomater.* **2015**, *24*, 12–23. [CrossRef] [PubMed]

4. Hoop, M.; Chen, X.Z.; Ferrari, A.; Mushtaq, F.; Ghazaryan, G.; Tervoort, T.; Poulikakos, D.; Nelson, B.; Pan, S. Ultrasound-mediated piezoelectric differentiation of neuron-like PC12 cells on PVDF membranes. *Sci. Rep.* **2017**, *7*, 1–8. [CrossRef] [PubMed]
5. Bhang, S.H.; Jang, W.S.; Han, J.; Yoon, J.K.; La, W.G.; Lee, E.; Kim, Y.S.; Shin, J.Y.; Lee, T.J.; Baik, H.K.; et al. Zinc oxide nanorod, based piezoelectric dermal patch for wound healing. *Adv. Funct. Mater.* **2017**, *27*, 1603497. [CrossRef]
6. Yin, Z.; Chen, X.; Chen, J.L.; Shen, W.L.; Nguyen, T.M.; Gao, L.; Ouyang, H.W. The regulation of tendon stem cell differentiation by the alignment of nanofibers. *Biomaterials* **2010**, *31*, 2163–2175. [CrossRef]
7. Kapat, K.; Shubhra, Q.T.; Zhou, M.; Leeuwenburgh, S. Piezoelectric nano-biomaterials for biomedicine and tissue regeneration. *Adv. Funct. Mater.* **2020**, *30*, 1909045. [CrossRef]
8. Kaliannagounder, V.K.; Raj, N.P.; Unnithan, A.R.; Park, J.; Park, S.S.; Kim, S.J.; Park, C.H.; Kim, C.S.; Sasikala, A.R. Remotely controlled self-powering electrical stimulators for osteogenic differentiation using bone inspired bioactive piezoelectric whitlockite nanoparticles. *Nano Energy* **2021**, *85*, 105901. [CrossRef]
9. Zaszczynska, A.; Sajkiewicz, P.; Gradys, A. Piezoelectric scaffolds as smart materials for neural tissue engineering. *Polymers* **2020**, *12*, 161. [CrossRef]
10. De, I.; Sharma, P.; Singh, M. Emerging approaches of neural regeneration using physical stimulations solely or coupled with smart piezoelectric nano-biomaterials. *Eur. J. Pharm. Biopharm.* **2022**, *173*, 73–91. [CrossRef]
11. Matichescu, A.; Ardelean, L.C.; Rusu, L.C.; Craciun, D.; Bratu, E.A.; Babucea, M.; Leretter, M. Advanced biomaterials and techniques for oral tissue engineering and regeneration-a review. *Materials* **2020**, *13*, 5303. [CrossRef] [PubMed]
12. Assimacopoulos, D. Wound healing promotion by the use of negative electric current. *Am. Surg.* **1968**, *34*, 423–431. [CrossRef] [PubMed]
13. Bottino, M.C.; Pankajakshan, D.; Nör, J.E. Advanced scaffolds for dental pulp and periodontal regeneration. *Dent. Clin.* **2017**, *61*, 689–711. [CrossRef]
14. Moussa, D.G.; Aparicio, C. Present and future of tissue engineering scaffolds for dentin, Äêpulp complex regeneration. *J. Tissue Eng. Regen. Med.* **2019**, *13*, 58–75. [CrossRef]
15. Pullar, C.E.; Baier, B.S.; Kariya, Y.; Russell, A.J.; Horst, B.A.; Marinkovich, M.P.; Isseroff, R.R. β4-Integrin and epidermal growth factor coordinately regulate electric field-mediated directional migration via Rac1. *Mol. Biol. Cell* **2006**, *17*, 4925–4935. [CrossRef]
16. Montoya, C.; Kurylec, J.; Baraniya, D.; Tripathi, A.; Puri, S.; Orrego, S. Antifungal Effect of Piezoelectric Charges on PMMA Dentures. *ACS Biomater. Sci. Eng.* **2021**, *7*, 4838–4846. [CrossRef]
17. Dhall, A.; Islam, S.; Park, M.; Zhang, Y.; Kim, A.; Hwang, G. Bimodal nanocomposite platform with antibiofilm and self-powering functionalities for biomedical applications. *ACS Appl. Mater. Interfaces* **2021**, *34*, 40379–40391. [CrossRef] [PubMed]
18. Chen, P.; Wang, H.; He, M.; Chen, B.; Yang, B.; Hu, B. Size-dependent cytotoxicity study of ZnO nanoparticles in HepG2 cells. *Ecotoxicol. Environ. Saf.* **2019**, *171*, 337–346. [CrossRef]
19. Fan, B.; Guo, Z.; Li, X.; Li, S.; Gao, P.; Xiao, X.; Wu, J.; Shen, C.; Jiao, Y.; Hou, W. Electroactive barium titanate coated titanium scaffold improves osteogenesis and osseointegration with low-intensity pulsed ultrasound for large segmental bone defects. *Bioact. Mater.* **2020**, *5*, 1087–1101. [CrossRef]
20. Fu, J.Y.; Liu, P.Y.; Cheng, J.; Bhalla, A.S.; Guo, R. Optical measurement of the converse piezoelectric d 33 coefficients of bulk and microtubular zinc oxide crystals. *Appl. Phys. Lett.* **2007**, *90*, 212907. [CrossRef]
21. Rasmussen, J.W.; Martinez, E.; Louka, P.; Wingett, D.G. Zinc oxide nanoparticles for selective destruction of tumor cells and potential for drug delivery applications. *Expert Opin. Drug Deliv.* **2010**, *7*, 1063–1077. [CrossRef] [PubMed]
22. Mould, R.F. Pierre Curie, 1859–1906. *Curr. Oncol.* **2007**, *14*, 74–82. [CrossRef] [PubMed]
23. Jaffe, H. Piezoelectric ceramics. *J. Am. Ceram. Soc.* **1958**, *41*, 494–498. [CrossRef]
24. Michel, K.H.; Verberck, B. Phonon dispersions and piezoelectricity in bulk and multilayers of hexagonal boron nitride. *Phys. Rev. B* **2011**, *83*, 15328. [CrossRef]
25. Kim, I.D.; Avrahami, Y.; Socci, L.; Lopez-Royo, F.; Tuller, H.L. Ridge waveguide using highly oriented BaTiO3 thin films for electro-optic application. *J. Asian Ceram. Soc.* **2014**, *3*, 231–234. [CrossRef]
26. Mokhtari, F.; Azimi, B.; Salehi, M.; Hashemikia, S.; Danti, S. Recent advances of polymer-based piezoelectric composites for biomedical applications. *J. Mech. Behav. Biomed. Mater.* **2021**, *122*, 104669. [CrossRef]
27. Yang, Z.; Zhou, S.; Zu, J.; Inman, D. High-performance piezoelectric energy harvesters and their applications. *Joule* **2018**, *2*, 642–697. [CrossRef]
28. Dahiya, R.S.; Valle, M. *Robotic Tactile Sensing: Technologies and System*; Springer: Berlin/Heidelberg, Germany, 2013; pp. 1–245. [CrossRef]
29. Cafarelli, A.; Marino, A.; Vannozzi, L.; Puigmart Luis, J.; Pan, S.; Ciofani, G.; Ricotti, L. Piezoelectric nanomaterials activated by ultrasound: The pathway from discovery to future clinical adoption. *ACS Nano* **2021**, *15*, 11066–11086. [CrossRef]
30. Krauskopf, E. A bibiliometric analysis of the Journal of Infection and Public Health: 2008–2016. *J. Infect. Public Health* **2018**, *11*, 224–229. [CrossRef]
31. Sweileh, W.M.; Al-Jabi, S.W.; Sa'ed, H.Z.; Sawalha, A.F.; Abu-Taha, A.S. Global research output in antimicrobial resistance among uropathogens: A bibliometric analysis (2002–2016). *J. Glob. Antimicrob. Resist.* **2018**, *13*, 104–114. [CrossRef]
32. Zhao, J.; Yu, G.; Cai, M.; Lei, X.; Yang, Y.; Wang, Q.; Zhai, X. Bibliometric analysis of global scientific activity on umbilical cord mesenchymal stem cells: A swiftly expanding and shifting focus. *Stem Cell Res. Ther.* **2018**, *9*, 32. [CrossRef] [PubMed]

33. Wang, X.; Sun, F.; Yin, G.; Wang, Y.; Liu, B.; Dong, M. Tactile-sensing based on flexible PVDF nanofibers via electrospinning: A review. *Sensors* **2018**, *18*, 330. [CrossRef] [PubMed]
34. Mirzaei, A.; Moghadam, A.S.; Abazari, M.F.; Nejati, F.; Torabinejad, S.; Kaabi, M.; Enderami, S.E.; Ardeshirylajimi, A.; Darvish, M.; Soleimanifar, F.; et al. Comparison of osteogenic differentiation potential of induced pluripotent stem cells on 2D and 3D polyvinylidene fluoride scaffolds. *J. Cell. Physiol.* **2019**, *234*, 17854–17862. [CrossRef] [PubMed]
35. Prescott, R.S.; Alsanea, R.; Fayad, M.I.; Johnson, B.R.; Wenckus, C.S.; Hao, J.; John, A.S.; George, A. In vivo generation of dental pulp-like tissue by using dental pulp stem cells, a collagen scaffold, and dentin matrix protein 1 after subcutaneous transplantation in mice. *J. Endod.* **2008**, *34*, 421–426. [CrossRef] [PubMed]
36. Wang, R.; Sui, J.; Wang, X. Natural Piezoelectric Biomaterials: A Biocompatible and Sustainable Building Block for Biomedical Devices. *ACS Nano* **2022**, *11*, 17708–17728. [CrossRef]
37. More, N.; Kapusetti, G. Piezoelectric material-a promising approach for bone and cartilage regeneration. *Med. Hypotheses* **2017**, *108*, 10–16. [CrossRef]
38. Jiang, S.; Yu, Z.; Zhang, L.; Wang, G.; Dai, X.; Lian, X.; Yan, Y.; Zhang, L.; Wang, Y.; Li, R.; et al. Effects of different aperture-sized type I collagen/silk fibroin scaffolds on the proliferation and differentiation of human dental pulp cells. *Regen. Biomater.* **2021**, *8*, rbab028. [CrossRef]
39. Yucel, T.; Cebe, P.; Kaplan, D.L. Structural origins of silk piezoelectricity. *Adv. Funct. Mater.* **2011**, *21*, 779–785. [CrossRef]
40. Uchino, K. The development of piezoelectric materials and the new perspective. In *Advanced Piezoelectric Materials*; Woodhead Publishing: Sawston, UK, 2017; pp. 1–92. [CrossRef]
41. Savakus, H.P.; Klicker, K.A.; Newnham, R.E. PZT-epoxy piezoelectric transducers: A simplified fabrication procedure. *Mater. Res. Bull.* **1981**, *16*, 677–680. [CrossRef]
42. Hu, Y.; Wang, Z.L. Recent progress in piezoelectric nanogenerators as a sustainable power source in self-powered systems and active sensors. *Nano Energy* **2015**, *14*, 3–14. [CrossRef]
43. Ciofani, G.; Ricotti, L.; Canale, C.; D'Alessandro, D.; Berrettini, S.; Mazzolai, B.; Mattoli, V. Effects of barium titanate nanoparticles on proliferation and differentiation of rat mesenchymal stem cells. *Colloids Surf. B Biointerfaces* **2013**, *102*, 312–320. [CrossRef] [PubMed]
44. Montoya, C.; Jain, A.; LondonÃÉo, J.J.; Correa, S.; Lelkes, P.I.; Melo, M.A.; Orrego, S. Multifunctional Dental Composite with Piezoelectric Nanofillers for Combined Antibacterial and Mineralization Effects. *ACS Appl. Mater. Interfaces* **2021**, *37*, 43868–43879. [CrossRef] [PubMed]
45. Liu, J.; Kang, Y.; Yin, S.; Song, B.; Wei, L.; Chen, L.; Shao, L. Zinc oxide nanoparticles induce toxic responses in human neuroblastoma SHSY5Y cells in a size-dependent manner. *Int. J. Nanomed.* **2017**, *12*, 8085. [CrossRef] [PubMed]
46. Ramasamy, M.; Das, M.; An, S.S.; Yi, D.K. Role of surface modification in zinc oxide nanoparticles and its toxicity assessment toward human dermal fibroblast cells. *Int. J. Nanomed.* **2014**, *9*, 3707. [CrossRef]
47. Yoon, J.K.; Misra, M.; Yu, S.J.; Kim, H.Y.; Bhang, S.H.; Song, S.Y.; Lee, J.R.; Ryu, S.; Choo, Y.W.; Jeong, G.J.; et al. Thermosensitive, stretchable, and piezoelectric substrate for generation of myogenic cell sheet fragments from human mesenchymal stem cells for skeletal muscle regeneration. *Adv. Funct. Mater.* **2017**, *27*, 1703853. [CrossRef]
48. Wang, J.; Liu, X.; Jin, X.; Ma, H.; Hu, J.; Ni, L.; Ma, P.X. The odontogenic differentiation of human dental pulp stem cells on nanofibrous poly (L-lactic acid) scaffolds in vitro and in vivo. *Acta Biomater.* **2010**, *6*, 3856–3863. [CrossRef]
49. Gao, Q.; Cooper, P.R.; Walmsley, A.D.; Scheven, B.A. Role of Piezo channels in ultrasound-stimulated dental stem cells. *J. Endod.* **2017**, *43*, 1130–1136. [CrossRef]
50. Lai, C.Y.; Groth, A.; Gray, S.; Duke, M. Nanocomposites for improved physical durability of porous PVDF membranes. *Membranes* **2014**, *4*, 55–78. [CrossRef]
51. Kalimuldina, G.; Turdakyn, N.; Abay, I.; Medeubayev, A.; Nurpeissova, A.; Adair, D.; Bakenov, Z. A review of piezoelectric PVDF film by electrospinning and its applications. *Sensors* **2020**, *20*, 5214. [CrossRef]
52. Martins, P.M.; Ribeiro, S.; Ribeiro, C.; Sencadas, V.; Gomes, A.C.; Gama, F.M.; Lanceros Sandez, S. Effect of poling state and morphology of piezoelectric poly (vinylidene fluoride) membranes for skeletal muscle tissue engineering. *Rsc Adv.* **2013**, *39*, 17938–17944. [CrossRef]
53. Ribeiro, C.; Prssinen, J.; Sencadas, V.; Correia, V.; Miettinen, S.; Hytnen, V.P.; Lanceros, Sandez, S. Dynamic piezoelectric stimulation enhances osteogenic differentiation of human adipose stem cells. *J. Biomed. Mater. Res. Part A* **2015**, *103*, 2172–2175. [CrossRef]
54. Szewczyk, P.K.; Metwally, S.; Karbowniczek, J.E.; Marzec, M.M.; Stodolak-Zych, E.; Gruszczyński, A.; Bernasik, A.; Stachewicz, U. Surface-potential-controlled cell proliferation and collagen mineralization on electrospun polyvinylidene fluoride (PVDF) fiber scaffolds for bone regeneration. *ACS Biomate. Sci. Eng.* **2018**, *5*, 582–593. [CrossRef] [PubMed]
55. Tai, Y.; Yang, S.; Yu, S.; Banerjee, A.; Myung, N.V.; Nam, J. Modulation of piezoelectric properties in electrospun PLLA nanofibers for application-specific self-powered stem cell culture platforms. *Nano Energy* **2021**, *89*, 106444. [CrossRef]
56. Mirzaei, A.; Saburi, E.; Enderami, S.E.; Barati Bagherabad, M.; Enderami, S.E.; Chokami, M.; Shapouri Moghadam, A.; Salarinia, R.; Ardeshirylajimi, A.; Mansouri, V.; et al. Synergistic effects of polyaniline and pulsed electromagnetic field to stem cells osteogenic differentiation on polyvinylidene fluoride scaffold. *Artif. Cells Nanomed. Biotechnol.* **2019**, *47*, 3058–3066. [CrossRef]
57. Dodds, J.S.; Meyers, F.N.; Loh, K.J. Piezoelectric characterization of PVDF-TrFE thin films enhanced with ZnO nanoparticles. *IEEE Sens. J.* **2011**, *12*, 1889–1890. [CrossRef]

58. Parssinen, J.; Hammarn, H.; Rahikainen, R.; Sencadas, V.; Ribeiro, C.; Vanhatupa, S.; Miettinen, S.; Lanceros Sandez, S.; Hytdnen, V.P. Enhancement of adhesion and promotion of osteogenic differentiation of human adipose stem cells by poled electroactive poly (vinylidene fluoride). *J. Biomed. Mater. Res. Part A* **2015**, *103*, 919–928. [CrossRef]
59. Linde, A. Dentin matrix proteins: Composition and possible functions in calcification. *Anat. Rec.* **1989**, *224*, 154–166. [CrossRef]
60. Huang, Y.; Rui, G.; Li, Q.; Allahyarov, E.; Li, R.; Fukuto, M.; Zhong, G.J.; Xu, J.Z.; Li, Z.M.; Taylor, P.L.; et al. Enhanced piezoelectricity from highly polarizable oriented amorphous fractions in biaxially oriented poly (vinylidene fluoride) with pure Œ ≤ crystals. *Nat. Commun.* **2021**, *12*, 675. [CrossRef]
61. Damaraju, S.M.; Shen, Y.; Elele, E.; Khusid, B.; Eshghinejad, A.; Li, J.; Jaffe, M.; Arinzeh, T.L. Three-dimensional piezoelectric fibrous scaffolds selectively promote mesenchymal stem cell differentiation. *Biomaterials* **2017**, *149*, 51–62. [CrossRef]
62. Park, M.; Islam, S.; Kim, H.E.; Korostoff, J.; Blatz, M.B.; Hwang, G.; Kim, A. Human oral motion, powered smart dental implant (SDI) for in situ ambulatory photobiomodulation therapy. *Adv. Healthc. Mater.* **2020**, *9*, 2000658. [CrossRef]
63. Rizk, A.; Rabie, A.B. Human dental pulp stem cells expressing transforming growth factor Œ≤3 transgene for cartilage-like tissue engineering. *Cytotherapy* **2013**, *15*, 712–725. [CrossRef] [PubMed]
64. Soares, D.G.; Zhang, Z.; Mohamed, F.; Eyster, T.W.; de Souza Costa, C.A.; Ma, P.X. Simvastatin and nanofibrous poly (l-lactic acid) scaffolds to promote the odontogenic potential of dental pulp cells in an inflammatory environment. *Acta Biomater.* **2018**, *68*, 190–203. [CrossRef] [PubMed]
65. Smith, M.; Calahorra, Y.; Jing, Q.; Kar-Narayan, S. Direct observation of shear piezoelectricity in poly-l-lactic acid nanowires. *APL Mater.* **2017**, *5*, 074105. [CrossRef]
66. Fukada, E. Piezoelectricity of biopolymers. *Biorheology* **1995**, *32*, 593–609. [CrossRef] [PubMed]
67. Chandrahasa, S.; Murray, P.E.; Namerow, K.N. Proliferation of mature ex vivo human dental pulp using tissue engineering scaffolds. *J. Endod.* **2011**, *37*, 1236–1239. [CrossRef]
68. Das, R.; Curry, E.J.; Le, T.T.; Awale, G.; Liu, Y.; Li, S.; Contreras, J.; Bednarz, C.; Millender, J.; Xin, X.; et al. Biodegradable nanofiber bone-tissue scaffold as remotely controlled and self-powering electrical stimulator. *Nano Energy* **2020**, *76*, 105028. [CrossRef]
69. Da Silva, D.; Kaduri, M.; Poley, M.; Adir, O.; Krinsky, N.; Shainsky-Roitman, J.; Schroeder, A. Biocompatibility, biodegradation and excretion of polylactic acid (PLA) in medical implants and theranostic systems. *Chem. Eng. J.* **2018**, *340*, 9–14. [CrossRef]
70. Esmaeili, M.; Baei, M.S. Fabrication of biodegradable polymer nanocomposite from copolymer synthesized by C. necator for bone tissue engineering. *World Appl. Sci. J.* **2011**, *14*, 106–111. Available online: https://citeseerx.ist.psu.edu/viewdoc/download?doi=10.1.1.567.9599&rep=rep1&type=pdf (accessed on 20 July 2022).
71. Kose, G.T.; Korkusuz, F.; Ozkul, A.; Soysal, Y.; Ozdemir, T.; Yildiz, C.; Hasirci, V. Tissue engineered cartilage on collagen and PHBV matrices. *Biomaterials* **2005**, *26*, 5187–5197. [CrossRef]
72. Jacob, J.; More, N.; Mounika, C.; Gondaliya, P.; Kalia, K.; Kapusetti, G. Smart piezoelectric nanohybrid of poly (3-hydroxybutyrate-co-3-hydroxyvalerate) and barium titanate for stimulated cartilage regeneration. *ACS Appl. Bio Mater.* **2019**, *2*, 4922–4931. [CrossRef]
73. Ferreira, A.M.; Gentile, P.; Chiono, V.; Ciardelli, G. Collagen for bone tissue regeneration. *Acta Biomater.* **2012**, *8*, 3191–3200. [CrossRef] [PubMed]
74. Silva, C.C.; Lima, C.G.; Pinheiro, A.G.; Goes, J.C.; Figueiro, S.D.; Sombra, A.S. On the piezoelectricity of collagen–chitosan films. *Phys. Chem. Chem. Phys.* **2001**, *3*, 4154–4157. [CrossRef]
75. Azzopardi, N.; Moharamzadeh, K.; Wood, D.J.; Martin, N.; van Noort, R. Effect of resin matrix composition on the translucency of experimental dental composite resins. *Dent. Mater.* **2009**, *25*, 1564–1568. [CrossRef] [PubMed]
76. Kwon, J.; Cho, H. Piezoelectric heterogeneity in collagen type I fibrils quantitatively characterized by piezoresponse force microscopy. *ACS Biomater. Sci. Eng.* **2020**, *6*, 6680–6689. [CrossRef] [PubMed]
77. Kitasako, Y.; Shibata, S.; Cox, C.F.; Tagami, J. Location, arrangement and possible function of interodontoblastic collagen fibres in association with calcium hydroxide-induced hard tissue bridges. *Int. Endod. J.* **2002**, *35*, 996–1004. [CrossRef] [PubMed]
78. Sumita, Y.; Honda, M.J.; Ohara, T.; Tsuchiya, S.; Sagara, H.; Kagami, H.; Ueda, M. Performance of collagen sponge as a 3-D scaffold for tooth-tissue engineering. *Biomaterials* **2006**, *27*, 3238–3248. [CrossRef]
79. Silva, C.C.; Thomazini, D.; Pinheiro, A.G.; Aranha, N.; Figueiro, S.D.; Ges, J.C.; Sombra, A.S. Collagen-hydroxyapatite films: Piezoelectric properties. *Mater. Sci. Eng. B* **2001**, *86*, 210–218. [CrossRef]
80. Yang, X.; Han, G.; Pang, X.; Fan, M. Chitosan/collagen scaffold containing bone morphogenetic protein, DNA supports dental pulp stem cell differentiation in vitro and in vivo. *J. Biomed. Mater. Res. Part A* **2020**, *108*, 2519–2526. [CrossRef]
81. Matsunaga, T.; Yanagiguchi, K.; Yamada, S.; Ohara, N.; Ikeda, T.; Hayashi, Y. Chitosan monomer promotes tissue regeneration on dental pulp wounds. *J. Biomed. Mater. Res. Part A Off. J. Soc. Biomater.* **2006**, *76*, 711–720. [CrossRef]
82. Liu, S.; Lau, C.S.; Liang, K.; Wen, F.; Teoh, S.H. Marine collagen scaffolds in tissue engineering. *Curr. Opin. Biotechnol.* **2022**, *74*, 92–103. [CrossRef]
83. Chen, H.; Fan, M. Chitosan/carboxymethyl cellulose polyelectrolyte complex scaffolds for pulp cells regeneration. *J. Bioact. Compat. Polym.* **2007**, *22*, 475–491. [CrossRef]
84. Demarco, F.F.; Casagrande, L.; Zhang, Z.; Dong, Z.; Tarquinio, S.B.; Zeitlin, B.D.; Shi, S.; Smith, A.J.; Nor, J.E. Effects of morphogen and scaffold porogen on the differentiation of dental pulp stem cells. *J. Endod.* **2010**, *36*, 1805–1811. [CrossRef] [PubMed]
85. Liao, F.; Chen, Y.; Li, Z.; Wang, Y.; Shi, B.; Gong, Z.; Cheng, X. A novel bioactive three-dimensional (tricalcium phosphate/chitosan scaffold for periodontal tissue engineering. *J. Mater. Sci. Mater. Med.* **2010**, *21*, 489–496. [CrossRef] [PubMed]

86. Pattanaik, S.; Jena, A.; Shashirekha, G. In vitro comparative evaluation of antifungal efficacy of three endodontic sealers with and without incorporation of chitosan nanoparticles against Candida albicans. *J. Conserv. Dent.* **2019**, *22*, 564. [CrossRef]
87. Shrestha, A.; Hamblin, M.R.; Kishen, A. Photoactivated rose bengal functionalized chitosan nanoparticles produce antibacterial/biofilm activity and stabilize dentin-collagen. *Nanomed. Nanotechnol. Biol. Med.* **2014**, *10*, 491–501. [CrossRef]
88. Praveen, E.; Murugan, S.; Jayakumar, K. Investigations on the existence of piezoelectric property of a bio-polymer-chitosan and its application in vibration sensors. *RSC Adv.* **2017**, *7*, 35490–35495. [CrossRef]
89. Hosseini, E.S.; Manjakkal, L.; Shakthivel, D.; Dahiya, R. Glycine–chitosan-based flexible biodegradable piezoelectric pressure sensor. *ACS App. Mate. Interfaces* **2020**, *12*, 9008–9016. [CrossRef]
90. Song, Y.; Shi, Z.; Hu, G.H.; Xiong, C.; Isogai, A.; Yang, Q. Recent advances in cellulose-based piezoelectric and triboelectric nanogenerators for energy harvesting: A review. *J. Mater. Chem. A* **2021**, *9*, 1910–1937. [CrossRef]
91. Lindman, B.; Karlstrodm, G.; Stigsson, L. On the mechanism of dissolution of cellulose. *J. Mol. Liq.* **2010**, *156*, 76–81. [CrossRef]
92. O'sullivan, A.C. Cellulose: The structure slowly unravels. *Cellulose* **1997**, *4*, 173–207. [CrossRef]
93. Zugenmaier, P. *Crystalline Cellulose and Derivatives: Characterization and Structures*; Springer Science & Business Media: Berlin/Heidelberg, Germany, 2008. [CrossRef]
94. VanderHart, D.L.; Atalla, R.H. Studies of microstructure in native celluloses using solid-state carbon-13 NMR. *Macromolecules* **1984**, *17*, 1465–1472. [CrossRef]
95. Klemm, D.; Heublein, B.; Fink, H.P.; Bohn, A. Cellulose: Fascinating biopolymer and sustainable raw material. *Angew. Chem. Int. Ed.* **2005**, *22*, 3358–3393. [CrossRef] [PubMed]
96. Fukada, E. Piezoelectricity of wood. *J. Phys. Soc. Jpn.* **1955**, *10*, 149–154. Available online: https://www.fpl.fs.fed.us/documnts/fplgtr/fpl_gtr212.pdf (accessed on 20 July 2022). [CrossRef]
97. Zhong, C. Industrial-scale production and applications of bacterial cellulose. *Front. Bioeng. Biotechnol.* **2020**, *22*, 605374. [CrossRef]
98. An, S.J.; Lee, S.H.; Huh, J.B.; Jeong, S.I.; Park, J.S.; Gwon, H.J.; Kang, E.S.; Jeong, C.M.; Lim, Y.M. Preparation and characterization of resorbable bacterial cellulose membranes treated by electron beam irradiation for guided bone regeneration. *Int. J. Mol. Sci.* **2017**, *18*, 2236. [CrossRef]
99. Voicu, G.; Jinga, S.I.; Drosu, B.G.; Busuioc, C. Improvement of silicate cement properties with bacterial cellulose powder addition for applications in dentistry. *Carbohydr. Polym.* **2017**, *174*, 160–170. [CrossRef]
100. Woloszyk, A.; Buschmann, J.; Waschkies, C.; Stadlinger, B.; Mitsiadis, T.A. Human dental pulp stem cells and gingival fibroblasts seeded into silk fibroin scaffolds have the same ability in attracting vessels. *Front. Physiol.* **2016**, *7*, 140. [CrossRef]
101. Kweon, H.; Lee, S.W.; Hahn, B.D.; Lee, Y.C.; Kim, S.G. Hydroxyapatite and silk combination-coated dental implants result in superior bone formation in the peri-implant area compared with hydroxyapatite and collagen combination-coated implants. *J. Oral Maxillofac. Surg.* **2014**, *72*, 1928–1936. [CrossRef]
102. Huang, J.; Wong, C.; George, A.; Kaplan, D.L. The effect of genetically engineered spider silk-dentin matrix protein 1 chimeric protein on hydroxyapatite nucleation. *Biomaterials* **2007**, *28*, 2358–2367. [CrossRef]
103. Inoue, S.; Tanaka, K.; Arisaka, F.; Kimura, S.; Ohtomo, K.; Mizuno, S. Silk fibroin of Bombyx mori is secreted, assembling a high molecular mass elementary unit consisting of H-chain, L-chain, and P25, with a 6: 6: 1 molar ratio. *J. Biol. Chem.* **2000**, *275*, 40517–40528. [CrossRef]
104. Ando, Y.; Okano, R.; Nishida, K.; Miyata, S.; Fukada, E. Piezoelectric and related properties of hydrated silk fibroin. *Rep. Prog. Polym. Phys. Jpn.* **1980**, *23*, 775–778.
105. Fukada, E. Piezoelectric properties of biological polymers. *Q. Rev. Biophys.* **1983**, *16*, 59–87. [CrossRef] [PubMed]
106. Kim, D.; Han, S.A.; Kim, J.H.; Lee, J.H.; Kim, S.W.; Lee, S.W. Biomolecular piezoelectric materials: From amino acids to living tissues. *Adv. Mater.* **2020**, *32*, 1906989. [CrossRef] [PubMed]
107. Guerin, S.; Stapleton, A.; Chovan, D.; Mouras, R.; Gleeson, M.; McKeown, C.; Noor, M.R.; Silien, C.; Rhen, F.M.; Kholkin, A.L.; et al. Control of piezoelectricity in amino acids by supramolecular packing. *Nat. Mater.* **2018**, *17*, 180–186. [CrossRef] [PubMed]
108. Guerin, S.; O'Donnell, J.; Haq, E.U.; McKeown, C.; Silien, C.; Rhen, F.M.; Soulimane, T.; Tofail, S.A.; Thompson, D. Racemic amino acid piezoelectric transducer. *Phys. Rev. Lett.* **2019**, *122*, 047701. [CrossRef] [PubMed]
109. Guerin, S.; Tofail, S.A.; Thompson, D. Longitudinal piezoelectricity in orthorhombic amino acid crystal films. *Cryst. Growth Des.* **2018**, *18*, 4844–4848. [CrossRef]
110. Nakhmanson, S.M.; Calzolari, A.; Meunier, V.; Bernholc, J.; Nardelli, M.B. Spontaneous polarization and piezoelectricity in boron nitride nanotubes. *Phys. Rev. B* **2003**, *67*, 235406. [CrossRef]
111. Guerin, S.; Tofail, S.A.; Thompson, D. Organic piezoelectric materials: Milestones and potential. *NPG Asia Mater.* **2019**, *11*, 10. [CrossRef]
112. Bishara, H.; Nagel, A.; Levanon, M.; Berger, S. Amino acids nanocrystals for piezoelectric detection of ultra-low mechanical pressure. *Mat. Sci. Eng. C* **2020**, *108*, 110468. [CrossRef]
113. Vasilescu, D.; Cornillon, R.; Mallet, G. Piezoelectric resonances in amino acids. *Nature* **1970**, *225*, 635. [CrossRef]
114. Ren, K.; West, J.E.; Michael Yu, S. Planar microphone based on piezoelectric electrospun poly (γ-benzyl-α, L-glutamate) nanofibers. *J. Acoust. Soc. Am.* **2014**, *135*, EL291–EL297. [CrossRef] [PubMed]
115. Megelski, S.; Stephens, J.S.; Chase, D.B.; Rabolt, J.F. Micro-and nanostructured surface morphology on electrospun polymer fibers. *Macromolecules* **2002**, *35*, 8456–8466. [CrossRef]

116. Bae, W.J.; Min, K.S.; Kim, J.J.; Kim, J.J.; Kim, H.W.; Kim, E.C. Odontogenic responses of human dental pulp cells to collagen/nanobioactive glass nanocomposites. *Dent. Mater.* **2012**, *28*, 1271–1279. [CrossRef]
117. Jiang, W.; Li, L.; Zhang, D.; Huang, S.; Jing, Z.; Wu, Y.; Zhao, Z.; Zhao, L.; Zhou, S. Incorporation of aligned PCL-PEG nanofibers into porous chitosan scaffolds improved the orientation of collagen fibers in regenerated periodontium. *Acta Biomater.* **2015**, *25*, 240–252. [CrossRef]
118. Jiang, X.C.; Ding, J.; Kumar, A. Polyurethane-poly (vinylidene fluoride)(PU-PVDF) thin film composite membranes for gas separation. *J. Membr. Sci.* **2008**, *323*, 371–378. [CrossRef]
119. Popa, L.; Ghica, M.V.; Dinu-Pirvu, C. *Hydrogels: Smart Materials for Biomedical Applications*; BoD-Books on Demand: Norderstedt, Germany, 2019.
120. Klotz, B.J.; Oosterhoff, L.A.; Utomo, L.; Lim, K.S.; Vallmajo ÄêMartin, Q.; Clevers, H.; Woodfield, T.B.; Rosenberg, A.J.; Malda, J.; Ehrbar, M.; et al. A versatile biosynthetic hydrogel platform for engineering of tissue analogues. *Adv. Healthc. Mater.* **2019**, *19*, 1900979. [CrossRef]
121. Guo, H.F.; Li, Z.S.; Dong, S.W.; Chen, W.J.; Deng, L.; Wang, Y.F.; Ying, D.J. Piezoelectric PU/PVDF electrospun scaffolds for wound healing applications. *Colloids Surf. B Biointerfaces* **2012**, *96*, 29–36. [CrossRef]
122. Siddiqui, Z.; Sarkar, B.; Kim, K.K.; Kadincesme, N.; Paul, R.; Kumar, A.; Kobayashi, Y.; Roy, A.; Choudhury, M.; Yang, J.; et al. Angiogenic hydrogels for dental pulp revascularization. *Acta Biomater.* **2021**, *126*, 109–118. [CrossRef]
123. Li, Y.; Fu, R.; Guan, Y.; Zhang, Z.; Yang, F.; Xiao, C.; Wang, Z.; Yu, P.; Hu, L.; Zhou, Z.; et al. Piezoelectric Hydrogel for Prophylaxis and Early Treatment of Pressure Injuries/Pressure Ulcers. *ACS Biomater. Sci. Eng.* **2022**, *8*, 3078–3086. [CrossRef]
124. Zhao, W.; Shi, Z.; Hu, S.; Yang, G.; Tian, H. Understanding piezoelectric characteristics of PHEMA-based hydrogel nanocomposites as soft self-powered electronics. *Adv. Compos. Hybrid Mater.* **2018**, *1*, 320–331. [CrossRef]
125. Dawood, A.; Marti, B.M.; Sauret-Jackson, V.; Darwood, A. 3D printing in dentistry. *Br. Dent. J.* **2015**, *219*, 521–529. [CrossRef]
126. Tian, Y.; Chen, C.; Xu, X.; Wang, J.; Hou, X.; Li, K.; Lu, X.; Shi, H.; Lee, E.S.; Jiang, H.B. A review of 3D printing in dentistry: Technologies, affecting factors, and applications. *Scanning* **2021**, *2021*, 9950131. [CrossRef]
127. Prasad, S.; Kader, N.A.; Sujatha, G.; Raj, T.; Patil, S. 3D printing in dentistry. *J. 3D Print. Med.* **2018**, *2*, 89–91. [CrossRef]
128. Park, C.H.; Rios, H.F.; Jin, Q.; Bland, M.E.; Flanagan, C.L.; Hollister, S.J.; Giannobile, W.V. Biomimetic hybrid scaffolds for engineering human tooth-ligament interfaces. *Biomaterials* **2010**, *31*, 5945–5952. [CrossRef]
129. Cholleti, E.R. A Review on 3D printing of piezoelectric materials. *IOP Conf. Ser. Mater. Sci. Eng.* **2018**, *1*, 012046. [CrossRef]
130. Lee, C.; Tarbutton, J.A. Electric poling-assisted additive manufacturing process for PVDF polymer-based piezoelectric device applications. *Smart Mater. Struct.* **2014**, *23*, 095044. [CrossRef]
131. Zelenovskiy, P.; Kornev, I.; Vasilev, S.; Kholkin, A. On the origin of the great rigidity of self-assembled diphenylalanine nanotubes. *Phys. Chem. Chem. Phys.* **2016**, *18*, 29681–29685. [CrossRef]
132. Kuang, X.; Roach, D.J.; Wu, J.; Hamel, C.M.; Ding, Z.; Wang, T.; Dunn, M.L.; Qi, H.J. Advances in 4D printing: Materials and applications. *Adv. Funct. Mater.* **2019**, *29*, 1805290. [CrossRef]
133. Orrego, S.; Chen, Z.; Krekora, U.; Hou, D.; Jeon, S.Y.; Pittman, M.; Montoya, C.; Chen, Y.; Kang, S.H. Bioinspired Materials with Self, Adaptable Mechanical Properties. *Adv. Mater.* **2020**, *32*, 1906970. [CrossRef]

Journal of
Functional
Biomaterials

MDPI

Article

Cytotoxicity and Efficacy in Debris and Smear Layer Removal of HOCl-Based Irrigating Solution: An In Vitro Study

Goda Bilvinaite [1], Ruta Zongolaviciute [1], Saulius Drukteinis [1,*], Virginija Bukelskiene [2] and Elisabetta Cotti [3]

1 Institute of Dentistry, Faculty of Medicine, Vilnius University, Zalgirio 115, LT-08217 Vilnius, Lithuania; goda.bilvinaite@gmail.com (G.B.); ruta757@gmail.com (R.Z.)
2 Department of Biological Models, Life Sciences Center, Institute of Biochemistry, Vilnius University, LT-10257 Vilnius, Lithuania; virginija.bukelskiene@bchi.vu.lt
3 Department of Conservative Dentistry and Endodontics, University of Cagliari, 09124 Cagliari, Italy; cottiendo@gmail.com
* Correspondence: saulius.drukteinis@mf.vu.lt; Tel.: +370-610-41808

Abstract: In the present study we evaluated the cytotoxicity of super-oxidized water on human gingival fibroblasts and its efficacy in debris and smear layer removal from root canal walls. Cultured gingival fibroblasts were exposed to super-oxidized water (Sterilox), which was diluted in Iscove's modified Dulbecco's medium (IMDM) at 30%, 40%, 50%, 60% and 70% concentrations. The control group was maintained in IMDM. The cell viability was evaluated by means of an MTT assay after incubation periods of 1 h, 2 h, 24 h and 48 h. Pathological cellular changes were also observed under fluorescence and phase contrast microscopes. The efficacy in debris and smear layer removal was evaluated in comparison to the conventional application of sodium hypochlorite (NaOCl) and ethylenediaminetetraacetic acid (EDTA). Forty maxillary premolars were randomly divided into two equal groups ($n = 20$) and shaped with ProTaper NEXT rotary instruments using Sterilox or NaOCl/EDTA for irrigation. Afterwards, roots were split longitudinally and examined under a scanning electron microscope. The results revealed that super-oxidized water and sterile distilled water have acceptable biological properties for endodontic applications at concentrations up to 50% ($p > 0.05$). Moreover, super-oxidized water is equally effective in debris and smear layer removal as compared to NaOCl/EDTA ($p > 0.05$).

Keywords: cytotoxicity; debris; human gingival fibroblasts; irrigation; root canal; smear layer; Sterilox; super-oxidized water

1. Introduction

Failure of endodontic treatment is mostly related to the remaining microorganisms and their products within the root canal system [1]. Attempts to reduce the microbial load below a critical threshold, which can be subsequently controlled by the immune system [2], usually rely on a combination of mechanical root canal preparation and chemical disinfection [3]. Root canal instrumentation aims to remove necrotic pulp tissues, mechanically disrupt the biofilm matrix and facilitate the flow of the irrigant within the entire length of the root canal [4]. However, endodontic treatment of immature permanent teeth may limit the removal of root dentin to a minimum, since the cessation of root development makes the tooth more susceptible to fracture [5]. In these clinical situations, the debridement of the root canal system is mainly achieved by chemical means [6], which may also require special safety precautions due to the open root apex. It has been reported that nearly half of endodontists experience an accidental irrigant extrusion into the periapical tissues at least once, and open root apices, either iatrogenic or anatomic, may impel the occurrence of irrigant extrusion [7]. Even though various irrigation and irrigant activation techniques have been proposed to avoid inadvertent irrigant extrusion [8], some limitations in terms of safety or cleaning effectiveness usually remain [9]. Therefore, the irrigating solution should be selected appropriately in these clinical cases and should exhibit a high efficacy in root canal disinfection and debridement, along with the absence of toxicity towards the periodontal tissues.

Sodium hypochlorite (NaOCl) is generally assumed to be a benchmark for root canal irrigation due to its exceptional antimicrobial activity, tissue-dissolving properties, low cost and wide availability [1]. Traditionally, the use of NaOCl is followed by the use of calcium chelating agents, such as ethylenediaminetetraacetic acid (EDTA), which dissolve the inorganic part of the smear layer and result in a wider range of open dentinal tubules, thus contributing to the superior debridement and disinfection of the root canal system [10]. The combined application of NaOCl and EDTA has been successfully used in root canal treatment for decades and it is uncertain whether more cost-effective irrigating solutions will be ever found [11]. However, cytotoxicity is a well-known shortcoming of NaOCl that may cause the rapid destruction of periodontal tissues and provoke acute pain, swelling and hematoma when extruded beyond the root canal system [12]. The inadvertent extrusion of NaOCl can be extremely deleterious in immature permanent teeth, leading to the compromised viability of stem cells, which typically reside in close proximity to the root apex and guide the root formation process jointly with Hertwig's epithelial root sheath [5,13]. Moreover, EDTA associated with NaOCl has proved to induce dentin erosion and adversely affect the dentin microstructure by altering the primary ratio of organic and inorganic components [14]. All these deleterious effects, leading to a significant decrease in dentin's elastic modulus and flexural strength, as well as potential cellular damage, may jeopardize the long-term prognosis of the tooth [5,15]. Therefore, super-oxidized, or electrochemically activated, water has been previously suggested as one of the potential alternatives for the conventional application of NaOCl and EDTA [16].

Super-oxidized water is typically produced via the electrolysis of dilute salt (NaCl), and contains a mixture of chlorine-based oxidants, such as hypochlorous acid (HOCl), which penetrate the lipid bilayer of the cell membrane and kill pathogens through chlorination or oxidation processes [17]. Previous studies have confirmed that electrochemically activated water exhibits broad-spectrum antimicrobial activity, effectively prevents biofilm formation and leads to a considerable reduction in microbial load, with no significant differences to NaOCl or chlorhexidine gluconate (CHX) [18,19]. These disinfection properties have made super-oxidized water widely used in medicine for wound care and scar management [18]. Successful treatment outcomes, obtained in wound healing [20], suggest that electrochemically activated water provides biocompatibility and favorable cellular responses, making this solution a particularly attractive option for root canal treatment or the management of endodontic complications, where a high risk of irrigant extrusion into surrounding periodontal tissues exists. However, more evidence-based data on super-oxidized water cytotoxicity are needed to confirm its safe biological responses at the cellular level.

Previous studies have shown that electrochemically activated water has little or no impact on the organic component of the dentin matrix [11] and causes fewer dentinal erosion as compared to EDTA [21]. However, the efficacy of super-oxidized water in debris and smear layer removal is still a subject of controversy that needs to be addressed [11,21]. Therefore, in the present study we aimed to evaluate both the potential cytotoxicity of super-oxidized water on cultured human gingival fibroblasts and its efficacy in debris and smear layer removal. The null hypothesis tested was that super-oxidized water is a safe and non-toxic irrigating solution, removing debris and the smear layer from root canal walls with no significant differences compared to the conventional application of NaOCl and EDTA.

2. Materials and Methods

2.1. Assessment of Cytotoxity

2.1.1. Cell Culture Preparation

Human gingival fibroblasts were grown from a connective tissue graft, which was obtained from a healthy patient undergoing a gingivectomy procedure in the premolar region. The grafted tissue (2–3 mm^3) was transported in Dulbecco's modified Eagle's medium (DMEM; Gibco, Grand Island, NY, USA) supplemented with 250 U/mL penicillin, 0.25 mg/mL streptomycin, 0.05 mg/mL gentamycin and 200 U/mL nystatin. The specimen was subsequently washed with a new portion of DMEM and transferred to Iscove's modified Dulbecco's medium (IMDM; Gibco, Grand Island, NY, USA), containing 250 U/mL penicillin, 0.25 mg/mL streptomycin and 20% fetal bovine serum (FBS; Gibco, Grand Island, NY, USA). The grafted tissue was minced under sterile conditions, seeded in 96-well plates and incubated at 37 °C in a 5% CO_2 atmosphere and 95% relative humidity. The IMDM was changed every 48 h for 2 weeks until the confluence of 80% was reached. Afterwards, the cell monolayer was rinsed in phosphate buffered saline (PBS) without calcium and magnesium, dissociated with 0.25% trypsin/EDTA solution, seeded to new 96-well plates and continued to cultivate. The cell culture of 8–10 passages was used for further analysis.

2.1.2. MTT Assay

Human gingival fibroblasts were seeded in 96-well plates at a density of 5×10^3 cells per 100 μL of IMDM. After the incubation period of 24 h, IMDM was replaced with 100 μL of test medium-Sterilox solution (Optident, Ilkley, UK) + IMDM. Sterilox solution was diluted in IMDM at 30%, 40%, 50%, 60% and 70% concentrations. The control group was maintained in IMDM. All specimens were incubated for 1 h, 2 h, 24 h and 48 h. After the specified time periods, the test medium was removed and 10 μL of 3-(4,5-dimethylthiazol-2-yl)-2,5-diphenyltetrazolium bromide (MTT) was added to each well. Cells with MTT were incubated for 4 h at 37 °C in a humidified 5% CO_2 atmosphere. The MTT was subsequently aspirated and formazan crystals, produced by dehydrogenases in viable cells, were dissolved in the ethanol. The intensity of the colored solution was measured using a Tecan Infinite 200 absorbance microplate reader (Tecan, Männedorf, Switzerland) at a 570 nm wavelength. Cell viability was evaluated proportionally to absorbance. The mean values obtained from the control group were considered 100% cell viability, and values obtained from experimental groups were expressed as percentages of viable cells proportionally to the control group.

2.1.3. Fluorescence and Phase Contrast Microscopy

Human gingival fibroblasts were incubated for 24 h in test mediums and then treated with 1 μL of dual fluorescent staining solution containing 100 μg/mL acridine orange (AO) and 100 μg/mL ethidium bromide (EB). The 10 μL suspensions of stained cells were placed on a clean microscope slide and covered with a coverslip. Apoptotic cells were visualized under an Eclipse TS100-F fluorescent microscope (Nikon, Tokyo, Japan) at ×200 magnification.

Additional microscopy for the direct identification of pathological cellular changes was performed using an Eclipse TS100-F microscope with a phase contrast set at ×100 magnification.

2.2. Assessment of Smear Layer and Debris Removal

2.2.1. Specimen Selection and Preparation

A total of 40 human maxillary premolars having one root canal and fully developed root apices were included in this study, under the approval of the local ethics committee (protocol no. EK-2). Teeth were extracted for reasons unrelated to the study and were stored in an isotonic saline solution at room temperature until use.

Standard endodontic access cavities were prepared with a high-speed Endo Access bur (Dentsply Sirona, Ballaiques, Switzerland) under copious water-cooling. The working length (WL) was determined by inserting a size 10 K-file (Dentsply Sirona, Ballaiques, Switzerland) into the root canal until the tip of the instrument reached the apical foramen and was visible under ×10 magnification (OPMI Pico, Carl Zeiss, Oberkochen, Germany). The WL was established 1 mm from the apical foramen. Afterwards, the root apices were covered with a small amount of sticky wax to prevent the overflow of irrigating solution beyond the apical foramen.

Root canal shaping was performed with rotary nickel-titanium instruments ProTaper NEXT (Dentsply Sirona, Ballaiques, Switzerland) at the established WL. Instruments were driven at the rotation speed of 300 rpm and the torque of 1 Ncm in the following sequence: X1 (17/0.04), X2 (25/0.06), X3 (30/0.07), X4 (40/0.06). After the use of each instrument, root canals were repeatedly irrigated with an open-ended 29-G NaviTip needle (Ultradent Products Inc., South Jordan, UT, USA), attached to disposable syringes. The type of irrigating solution varied between two randomly allocated (www.random.org (accessed on 13 July 2022)) experimental groups (20 teeth per group):

- Sterilox group—root canals were repeatedly irrigated with 2 mL Sterilox solution, containing 200 ppm of available free chlorine. At the end of instrumentation, the irrigation needle was placed 1 mm from the WL and a final 1 min rinse with 4 mL Sterilox, followed by 4 mL distilled water, was performed, moving the needle along the root canal in a 5 mm amplitude.
- NaOCl/EDTA group—root canals were irrigated with 2 mL 3% NaOCl solution (Ultradent Products Inc., South Jordan, UT, USA) after every change of the instrument. The final 1 min flush was performed in the following sequence: 2 mL 3% NaOCl, 2 mL 18% EDTA (Ultradent Products Inc., South Jordan, UT, USA) and 4 mL distilled water.

At the end of the irrigation process, all root canals were dried with sterile paper points. Teeth were prepared by the same operator—an experienced endodontist.

2.2.2. Scanning Electron Microscopy

All specimens were decoronated at the cemento-enamel junction with a diamond-coated high-speed fissure bur and then grooved longitudinally on the buccal and lingual surfaces without penetrating the root canal. Roots were gently split into two halves using a small chisel, and completely dehydrated in a graded ethanol series at room temperature. The examination of each root half was performed under a Stereoscan 100 scanning electron microscope (Cambridge Instrument CO, Cambridge, UK). Ten microscopic fields at ×200 magnification (for debris) and fifteen at ×1000 (for smear layer) were assessed in the apical, middle and coronal thirds. The grading system was used to score the amount of superficial debris as follows:

- Score 1—clean root canal wall, only few small particles of debris;
- Score 2—small agglomerations of debris;
- Score 3—debris covering < 50% of the surface;
- Score 4—debris covering > 50% of the surface;
- Score 5—complete or nearly complete coverage by debris.

The presence of the smear layer was graded as follows:

- Score 1—no smear layer, dentinal tubules are open;
- Score 2—small amount of smear layer, some dentinal tubules are open;
- Score 3—homogenous smear layer covering the surface, only a few dentinal tubules are open;
- Score 4—complete coverage by homogenous smear layer, no open dentinal tubules;
- Score 5—complete coverage by a heavy, non-homogenous smear layer.

The scoring procedure was performed blindly by two trained and independent examiners, who scored each microscopic field from 1 to 5. The mean values of the debris and smear layer were calculated for each root canal third.

2.3. Statistical Analysis

Statistical analysis was performed using RStudio software (RStudio Inc., Boston, MA, USA). The assumption of the normality of cell viability data was assessed with the Shapiro–Wilk test and then the homogeneity of variance was confirmed via Levene's test. Statistically significant differences between data sets were determined using Student's *t*-test and one-way analysis of variance (ANOVA).

The smear layer and debris scores revealed a non-normal distribution, according to the Shapiro–Wilk test. Therefore, a non-parametric Mann–Whitney test was used for inter-group comparisons, and the Friedman test, followed by the Wilcoxon test, was selected for intra-group comparisons.

The significance level for all comparisons was set at 5%.

3. Results

3.1. Cytotoxity

The MTT assay revealed a tendency of Sterilox solutions to reduce cell viability in a concentration- and time-dependent manner. The lowest percentages of viable fibroblasts were observed after 24 h and 48 h incubation at a 70% concentration (Figure 1). No statistically significant differences were detected when the Sterilox and control groups were compared up to 50% concentrations. However, Sterilox was considerably more cytotoxic at concentrations above 50% after 24 h and 48 h ($p < 0.05$).

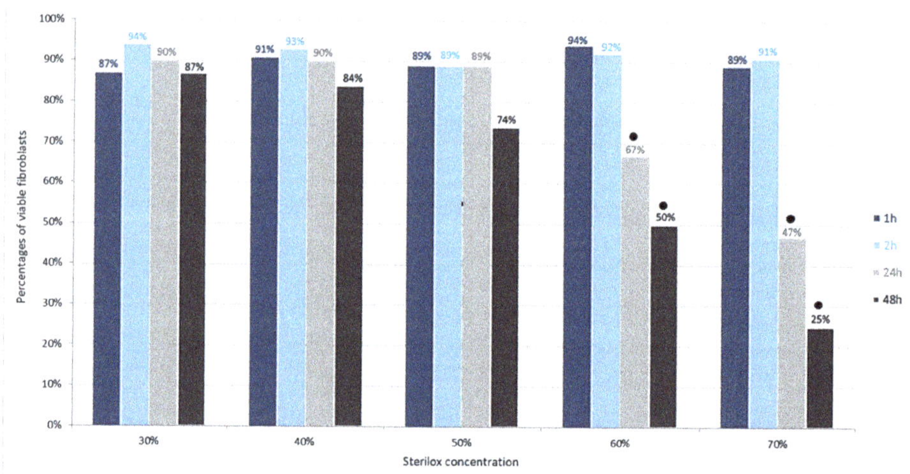

Figure 1. Percentages of viable human gingival fibroblasts after exposure to Sterilox. The black dots (•) above columns indicate statistically significant differences between Sterilox and control groups ($p < 0.05$).

The analysis of cell viability was supported by images obtained via fluorescence and phase contrast microscopy (Figure 2). Control cells demonstrated the typical fibroblast morphology, forming a sound monolayer, whereas exposure to Sterilox apparently induced a variety of pathological cellular changes. Apoptotic features at higher Sterilox concentrations appeared as cytoplasmic vacuolization, nuclear shrinkage (pyknosis) and cell rounding.

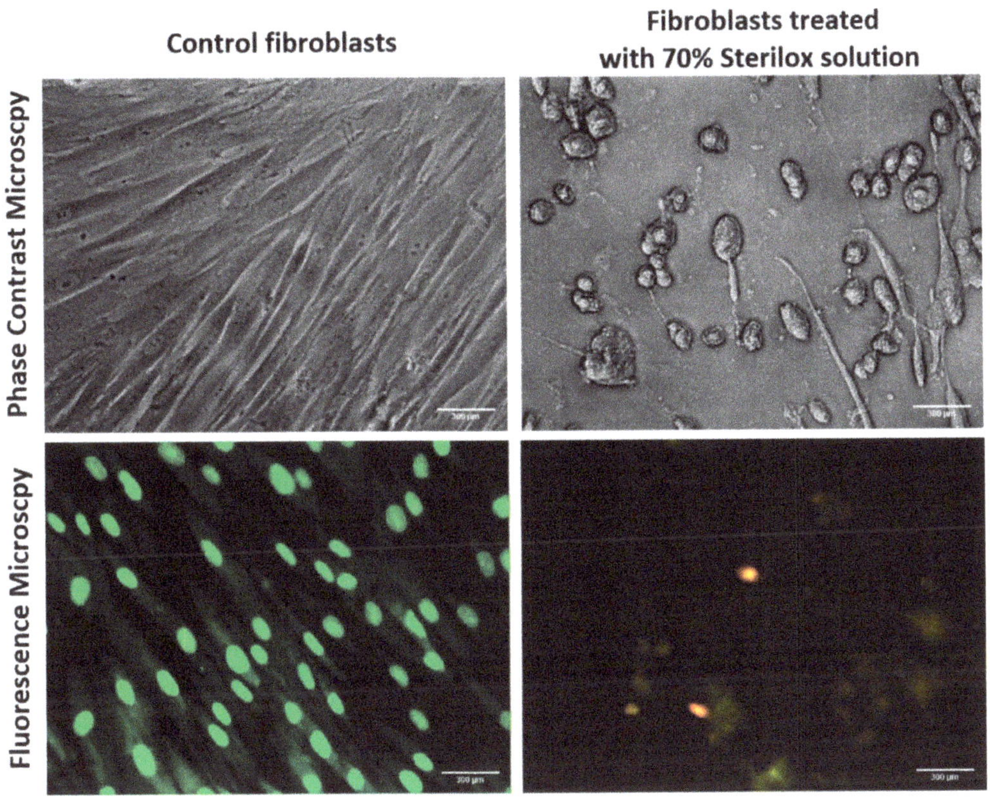

Figure 2. Representative microscopic images of human gingival fibroblasts at ×200 magnification after a 24 h incubation period. The bright red fluorescent coloration indicates the presence of apoptotic cells, whereas healthy viable cells appear green. Scale bar: 300 µm.

3.2. Smear Layer and Debris Removal

The mean scores of the debris and smear layer are summarized in Table 1. Even though some microscopic fields were free of a smear layer and debris (Figure 3), none of the irrigating solutions used in the present study was able to provide complete cleanliness along the entire length of the root canal.

Table 1. Mean scores and standard deviations of the debris and smear layer for each group in the different root canal third.

Group	Debris (n = 200)			Smear Layer (n = 300)		
	Coronal Third	Middle Third	Apical Third	Coronal Third	Middle Third	Apical Third
Sterilox	1.42 ± 0.42^{A}	1.99 ± 0.84	2.17 ± 1.06^{A}	2.55 ± 1.19^{B}	3.36 ± 1.22	3.67 ± 0.84^{B}
NaOCl/EDTA	1.41 ± 0.24^{C}	1.67 ± 0.72	1.83 ± 0.54^{C}	2.08 ± 1.11^{D}	2.53 ± 1.22	2.75 ± 0.93^{D}

n refers to the number of SEM photographs obtained and scored per group in each third. The same superscript letter indicates a significant difference between root canal thirds ($p < 0.05$).

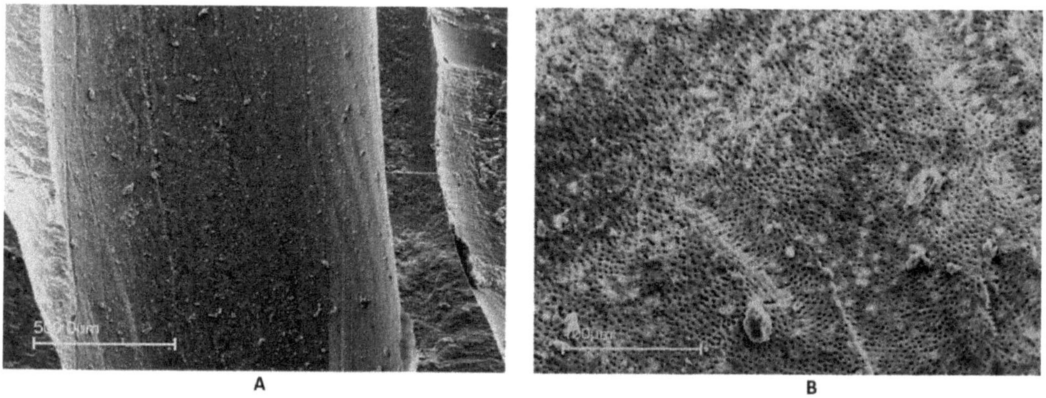

Figure 3. Representative SEM images of (A) debris-free root canal surface (middle third) at ×200 magnification; (B) smear layer-free root canal surface (coronal third) at ×1000 magnification.

The amount of smear layer and debris increased from the coronal to the apical third in both experimental groups. However, no statistically significant differences were detected between the Sterilox and NaOCl/EDTA groups ($p > 0.05$) regardless of the lower mean scores observed in the NaOCl/EDTA.

4. Discussion

The replacement of NaOCl and EDTA with a non-toxic and less erosive irrigating solution has been regarded as one of the preventive measures to avoid potential damage to periapical tissues and the dentin microstructure [22]. However, a preferred alternative to NaOCl and EDTA has not been identified to date. Even though CHX is usually suggested as an appropriate choice, it has been reported that CHX has little or no tissue-dissolving properties and its concentration recommended for endodontic treatment provides the cytotoxicity similar to that of NaOCl [7]. Moreover, if NaOCl is used to irrigate the root canal, the addition of CHX may lead to brown cytotoxic precipitate formation, with consequent tooth discoloration and a possible negative effect on the sealing ability of theobturation material [23]. Therefore, in the present study we focused on super-oxidized water, which has been previously suggested as another possible endodontic irrigant [24], belonging to the same group of chlorine-based solutions as NaOCl and CHX.

The main active compound of super-oxidized water is hypochlorous acid (HOCl), which dissociates to hypochlorite ions (OCl^-) only in alkaline environments and has extensively-studied antimicrobial activity against bacteria, fungi and viruses [18,25]. In contrast with hypochlorite ions (OCl^-), which dominate in NaOCl solution due to its high pH, the HOCl molecule is electrically neutral and can easily penetrate the target cell to exert a strong and rapid bactericidal response [17]. The detrimental effect of HOCl, triggering a variety of cell death mechanisms, has also been detected on different human cell lines [26]. However, evidence suggests that HOCl exhibits its cytotoxicity on human cells in a concentration-dependent manner [27], and only high local concentrations are associated with HOCl-induced cellular damage and oxidative stress [26]. These observations are in accordance with the present study, in which the apoptotic features of super-oxidized water, having an oxidation reduction potential of more than 950 mV, were directly related to the concentration.

The concentration with which HOCl induces irreversible damage to cells may also depend on the cell type [26]. However, the cytotoxicity of HOCl on cultured human gingival fibroblasts has not been investigated previously and thus the mechanism associated with fibroblast resistance to HOCl-mediated oxidative stress remains unclear. It is known that HOCl is a naturally occurring molecule produced by the immune system to destroy invading microorganisms [18]. During the activation of neutrophils, the respiratory burst generates hydrogen peroxide (H_2O_2) and subsequently the released enzyme myeloperoxidase catalyzes the formation of HOCl in the presence of H_2O_2 [28]. Therefore, it can be speculated that fibroblasts, which play an important role in the process of local inflammation [29], presumably possess a specific defense mechanism providing tolerance to a certain level of HOCl. This has been partially confirmed in a previous study, demonstrating that human gingival fibroblasts at the early stage of inflammation may increase the expression of special enzymes, which repair DNA damage caused by oxidative stress and provide an anti-apoptotic effect for up to 48 h [30]. This short-term self-recovery mechanism may explain the results of the present study, in which the viability of human gingival fibroblasts diminished in a time-dependent manner and reached its lowest level after the incubation period of 48 h. However, a significant decrease in cell viability was observed only with highly concentrated super-oxidized water, whereas concentrations up to 50% successfully retained acceptable biological properties within the specified time periods. These findings partially confirm the null hypothesis associated with the biocompatibility of super-oxidized water and may serve as a guideline for further clinical investigations, which are necessary to confirm the range of clinically safe concentrations, as various cellular processes that are unreproducible by in vitro models may also influence the quantitative extension of HOCl-induced cytotoxicity.

The ability to remove debris and the smear layer from root canal walls is another crucial property for irrigating solutions. Even though it is usually not considered to be of primary importance, non-removed accumulated hard tissue remnants may conceal microorganisms and act as a barrier, preventing the penetration of the irrigant and subsequently the sealer into the dentinal tubules [31]. The results of the present study demonstrated that super-oxidized water has no significant differences in debris and smear layer removal as compared to the conventional root canal irrigation protocol using NaOCl and EDTA. The exact mechanism by which super-oxidized water dissolves tissues is not fully understood. Evidence suggests that the tissue-dissolving properties of chlorine-based solutions are mainly determined by OCl^- ions, which begin to be released in low amounts from HOCl at pH values above 5.5 and become the completely dominant form at a pH of 9 [25,32]. However, the OCl^- concentration in super-oxidized water is generally assumed to remain low due to its pH range of 5 to 6.5 [24]. Theoretically, these pH values could be increased by the buffering effect of dentin, leading to the higher amount of OCl^- ions and thus the more efficient removal of debris and the smear layer [32,33]. Nevertheless, the influence of the dentin-buffering effect on super-oxidized water pH levels has not been investigated to date, and there is also a possibility that a low concentration of OCl^- might be compensated for by various other reactive ions and compounds generated during the electrolysis process.

Some limitations of SEM analysis should be also highlighted as one of the factors influencing the results of the present study. Even though SEM is a widely accepted method to assess the ability of endodontic irrigants to remove debris and smear layers [34], this method requires sample sectioning and allows the evaluation only of small selected areas, hence resulting in the potential loss of some valuable information [35]. In order to minimize these limitations, each third of gently halved roots was fully screened under SEM and only areas with the greatest amount of debris and smear layers were selected for further analysis. This evaluation method demonstrated that none of the irrigating solutions used in the present study was able to completely debride the root canal walls, even though some areas were proved to be free of smear layers and debris. The highest amount of residual debris and smear layer was observed in the apical root canal third for both experimental groups. These outcomes might be explained by the limited replenishment and exchange of the irrigating solution in the apical area, despite the sufficient root canal enlargement in the full working length [36]. Previous studies have demonstrated that the mechanical effect of the irrigation process has also a crucial role in the efficacy of root canal debridement [3], and thus various irrigation protocols following sonic or ultrasonic activation of the irrigating solution are highly recommended in terms of debris and smear layer removal [9]. However, the irrigation process in the present study was performed using syringes alone, with no variation in the volume and flow rate of the irrigants, with the aim of assessing the chemical debridement of root canals.

The current findings suggest that super-oxidized water, due to its relative non-toxicity and favorable efficacy in debris and smear layer removal, might be a noteworthy alternative to the conventional application of NaOCl and EDTA. It is not likely that super-oxidized water will ever replace NaOCl and EDTA in daily clinical practice, as it exerts no pulp tissue-dissolving properties [9], retains stability only for 14 days under ideal conditions [11] and contains a lower concentration of available free chlorine than NaOCl, thus requiring a larger amount of the irrigant for adequate disinfection [37]. However, in clinical situations, where irrigant extrusion is a great concern, super-oxidized water could be regarded as a safe and effective option. Further studies, especially well-designed clinical trials, would be highly valuable in order to confirm this hypothesis, as it is impossible to fully reproduce all the clinical conditions and possible biological reactions by means of simplified in vitro models.

5. Conclusions

Within the limitations of the present study, it can be concluded that super-oxidized water exhibits an acceptable biocompatibility at concentrations up to 50%, whereas higher concentrations exert apoptotic effects in a time-dependent manner. The efficacy of super-oxidized water in debris and smear layer removal exhibited no significant differences compared to the conventional irrigation protocol of NaOCl/EDTA, even though the superior debridement of root canals was observed in the NaOCl/EDTA group.

Author Contributions: Conceptualization, S.D. and V.B.; methodology, S.D and V.B.; software, R.Z.; validation, S.D., R.Z. and V.B.; formal analysis, R.Z.; investigation, S.D., R.Z. and V.B.; resources, V.B.; data curation, G.B.; writing—original draft preparation, G.B. and S.D.; writing—review and editing, E.C. and V.B.; visualization, G.B.; supervision, E.C.; project administration, S.D. All authors have read and agreed to the published version of the manuscript.

Funding: This research received no external funding.

Institutional Review Board Statement: The study was conducted according to the guidelines of the Declaration of Helsinki and approved by the Ethics Committee of Vilnius University Hospital Zalgirio Clinics (protocol no. EK-2, 2016 06 03).

Informed Consent Statement: Not applicable.

Data Availability Statement: Data are contained within the article.

Conflicts of Interest: The authors declare no conflict of interest. The sponsors had no role in the design of the study; in the collection, analysis or interpretation of data; in the writing of the manuscript or in the decision to publish the results.

References

1. Matos, F.S.; Khoury, R.D.; Carvalho, C.A.T.; Martinho, F.C.; Bresciani, E.; Valera, M.C. Effect of EDTA and QMIX ultrasonic activation on the reduction of microorganisms and endotoxins in ex vivo human root canals. *Braz. Dent. J.* **2019**, *30*, 220–226. [CrossRef] [PubMed]
2. Wenzler, J.S.; Falk, W.; Frankenberger, R.; Braun, A. Impact of adjunctive laser irradiation on the bacterial load of dental root canals: A randomized controlled clinical trial. *Antibiotics* **2021**, *10*, 1557. [CrossRef]
3. Versiani, M.A.; Alves, F.R.F.; Andrade-Junior, C.V.; Marceliano-Alves, M.F.; Provenzano, J.C.; Rôças, I.N.; Sousa-Neto, M.D.; Siqueira Jr, J.F. Micro-CT evaluation of the efficacy of hard-tissue removal from the root canal and isthmus area by positive and negative pressure irrigation systems. *Int. Endod. J.* **2016**, *49*, 1079–1087. [CrossRef] [PubMed]
4. Nagendrababu, V.; Jayaraman, J.; Suresh, A.; Kalyanasundaram, S.; Neelakantan, P. Effectiveness of ultrasonically activated irrigation on root canal disinfection: A systematic review of in vitro studies. *Clin. Oral. Investig.* **2018**, *22*, 655–670. [CrossRef]
5. Tavares, S.; Pintor, A.; de Almeida Barros Mourão, C.F.; Magno, M.; Montemezzi, P.; Sacco, R.; Alves, G.; Scelza, M.Z. Effect of different root canal irrigant solutions on the release of dentin-growth factors: A systematic review and meta-analysis. *Materials* **2021**, *14*, 5829. [CrossRef]
6. Shabahang, S. Treatment options: Apexogenesis and apexification. *J. Endod.* **2013**, *39*, 26–29. [CrossRef]
7. Guivarc, M.; Ordioni, U.; Mohamed Aly Ahmed, H.; Cohen, S.; Catherine, J.H.; Bukiet, F. Sodium hypochlorite accident: A systematic review. *J. Endod.* **2017**, *43*, 16–24. [CrossRef]

8. Iandolo, A.; Amato, A.; Pantaleo, G.; Dagna, A.; Ivaldi, L.; di Spirito, F.; Abdellatif, D. An innovative technique to safely perform active cleaning in teeth with open apices: CAB technique. *J. Conserv. Dent.* **2021**, *24*, 153–157. [CrossRef]
9. Boutsioukis, C.; Arias-Moliz, M.T. Irrigating Solutions, Devices, and Techniques. In *Endodontic Materials in Clinical Practice*, 1st ed.; Camilleri, J., Ed.; Wiley: Hoboken, NJ, USA, 2021; pp. 133–180.
10. Fernandes Zancan, R.; Hadis, M.; Burgess, D.; Zhang, Z.J.; Di Maio, A.; Tomson, P.; Hungaro Duarte, M.A.; Camilleri, J. A matched irrigation and obturation strategy for root canal therapy. *Sci. Rep.* **2021**, *11*, 4666. [CrossRef]
11. Ghisi, A.C.; Kopper, P.M.P.; Baldasso, F.E.R.; Stürmer, C.P.; Rossi-Fedele, G.; Steier, L.; de Figueiredo, J.A.P.; Morgental, R.D.; Vier-Pelisser, F.V. Effect of superoxidized water and sodium hypochlorite, associated or not with EDTA, on organic and inorganic components of bovine root dentin. *J. Endod.* **2015**, *41*, 925–930. [CrossRef]
12. Coaguila-Llerena, H.; Denegri-Hacking, A.; Lucano-Tinoco, L.; Mendiola-Aquino, C.; Faria, G. Accidental extrusion of sodium hypochlorite in a patient taking alendronate: A case report with an 8-year follow-up. *J. Endod.* **2021**, *47*, 1947–1952. [CrossRef]
13. Scott, M.B.; Zilinski, G.S.; Kirkpatrick, T.C.; Himel, V.T.; Sabey, K.A.; Lallier, T.E. The effects of irrigants on the survival of human stem cells of the apical papilla, including endocyn. *J. Endod.* **2018**, *44*, 263–268. [CrossRef] [PubMed]
14. Gawdat, S.I.; Bedier, M.M. Influence of dual rinse irrigation on dentinal penetration of a bioceramic root canal sealer: A Conofocal microscopic Analysis. *Aust. Endod. J.* **2021**. [CrossRef] [PubMed]
15. Kahler, B.; Rossi-Fedele, G.; Chugal, N.; Lin, L.M. An evidence-based review of the efficacy of treatment approaches for immature permanent teeth with pulp necrosis. *J. Endod.* **2017**, *43*, 1052–1057. [CrossRef] [PubMed]
16. Hsieh, S.C.; Teng, N.C.; Chu, C.C.; Chu, Y.T.; Chen, C.H.; Chang, L.Y.; Hsu, C.H.; Huang, C.S.; Hsiao, G.Y.W.; Yang, J.C. The antibacterial efficacy and in vivo toxicity of sodium hypochlorite and electrolyzed oxidizing (EO) water-based endodontic irrigating solutions. *Materials* **2020**, *13*, 260. [CrossRef]
17. Yan, P.; Daliri, E.B.M.; Oh, D.H. New clinical applications of electrolyzed water: A review. *Microorganisms* **2021**, *9*, 136. [CrossRef]
18. Gold, M.H.; Andriessen, A.; Bhatia, A.C.; Bitter, P.; Chilukuri, S.; Cohen, J.L.; Robb, C.W. Topical stabilized hypochlorous acid: The future gold standard for wound care and scar management in dermatologic and plastic surgery procedures. *J. Cosmet. Dermatol.* **2020**, *19*, 270–277. [CrossRef]
19. Giles, L.H.; Chan, S.N.; Jarad, F.; Hope, C. The efficacy of different endodontic irrigants against *Enterococcus faecalis* in an extracted human tooth model. *Int. Endod. J.* **2010**, *43*, 353. [CrossRef]
20. Weigelt, M.A.; McNamara, S.A.; Sanchez, D.; Hirt, P.A.; Kirsner, R.S. Evidence-based review of antibiofilm agents for wound care. *Adv. Wound Care* **2021**, *10*, 13. [CrossRef]
21. Rathakrishnan, M.; Sukumaran, V.G.; Subbiya, A. To evaluate the efficacy of an innovative irrigant on smear layer removal—SEM analysis. *J. Clin. Diagnostic. Res.* **2016**, *10*, 104–106. [CrossRef]
22. Boutsioukis, C.; Psimma, Z.; Van der Sluis, L.W.M. Factors affecting irrigant extrusion during root canal irrigation: A systematic review. *Int. Endod. J.* **2013**, *46*, 599–618. [CrossRef] [PubMed]
23. Nocca, G.; Ahmed, H.M.A.; Martorana, G.E.; Callà, C.; Gambarini, G.; Rengo, S.; Spagnuolo, G. Chromographic analysis and cytotoxic effects of chlorhexidine and sodium hypochlorite reaction mixtures. *J. Endod.* **2017**, *43*, 1545–1552. [CrossRef] [PubMed]
24. Hope, C.K.; Garton, S.G.; Wang, Q.; Burnside, G.; Farrelly, P.J. A direct comparison between extracted tooth and filter-membrane biofilm models of endodontic irrigation using *Enterococcus faecalis*. *Arch. Microbiol.* **2010**, *192*, 775–781. [CrossRef]
25. Gray, N.F. Free and combined chlorine. In *Microbiology of Waterborne Diseases: Microbiological Aspects and Risks*, 2nd ed.; Percival, S.T., Yates, M.V., Williams, D.D., Chalmers, R., Gray, N., Eds.; Elsevier: San Diego, CA, USA, 2014; pp. 571–590.
26. Yang, Y.T.; Whiteman, M.; Gieseg, S.P. HOCl causes necrotic cell death in human monocyte derived macrophages through calcium dependent calpain activation. *Biochim. Biophys. Acta* **2012**, *1823*, 420–429. [CrossRef] [PubMed]
27. Carmona-Gutierrez, D.; Alavian-Ghavanini, A.; Habernig, L.; Bauer, M.A.; Hammer, A.; Rossmann, C.; Zimmermann, A.; Ruckenstuhl, C.; Buttner, S.; Eisenberg, T.; et al. The cell death protease Kex1p is essential for hypochlorite-induced apoptosis in yeast. *Cell Cycle* **2013**, *12*, 1704–1712. [CrossRef] [PubMed]
28. Sakarya, S.; Gunay, N.; Karakulak, M.; Ozturk, B.; Ertugrul, B. Hypochlorous Acid: An ideal wound care agent with powerful microbicidal, antibiofilm, and wound healing potency. *Wounds Compend. Clin. Res. Pract.* **2014**, *26*, 342–350.
29. Zhang, L.; Wang, T.; Lu, Y.; Zheng, Q.; Gao, Y.; Zhou, X.; Huang, D. Tumor necrosis factor receptor-associated factor 6 plays a role in the inflammatory responses of human periodontal ligament fibroblasts to *Enterococcus faecalis*. *J. Endod.* **2015**, *41*, 1997–2001. [CrossRef]
30. Cheng, R.; Choudhury, D.; Liu, C.; Billet, S.; Hu, T.; Bhowmick, N. Gingival fibroblasts resist apoptosis in response to oxidative stress in a model of periodontal diseases. *Cell Death Discov.* **2015**, *1*, 1504. [CrossRef]
31. Lopes, R.M.V.; Marins, F.C.; Belladonna, F.G.; Souza, E.M.; De-Deus, G.; Lopes, R.T.; Silva, E.J.N.L. Untouched canal areas and debris accumulation after root canal preparation with rotary and adaptive systems. *Aust. Endod. J.* **2018**, *44*, 260–266. [CrossRef]
32. Macedo, R.G.; Herrero, N.P.; Wesselink, P.; Versluis, M.; Van Der Sluis, L. Influence of the dentinal wall on the pH of sodium hypochlorite during root canal irrigation. *J. Endod.* **2014**, *40*, 1005–1008. [CrossRef]
33. Jungbluth, H.; Marending, M.; De-Deus, G.; Sener, B.; Zehnder, M. Stabilizing sodium hypochlorite at high pH: Effects on soft tissue and dentin. *J. Endod.* **2011**, *37*, 693–696. [CrossRef] [PubMed]
34. Tonini, R.; Giovarruscio, M.; Gorni, F.; Ionescu, A.; Brambilla, E.; Irina, M.M.; Luzi, A.; Pires, P.M.; Sauro, S. In vitro evaluation of antibacterial properties and smear layer removal/sealer penetration of a novel silver-citrate root canal irrigant. *Materials* **2020**, *13*, 194. [CrossRef] [PubMed]

35. Plotino, G.; Colangeli, M.; Özyürek, T.; DeDeus, G.; Panzetta, C.; Castagnola, R.; Grande, N.M.; Marigo, L. Evaluation of smear layer and debris removal by stepwise intraoperative activation (SIA) of sodium hypochlorite. *Clin. Oral. Investig.* **2021**, *25*, 237–245. [CrossRef] [PubMed]
36. Miguéns-Vila, R.; Martín-Biedma, B.; Aboy-Pazos, S.; Uroz-Torres, D.; Álvarez-Nóvoa, P.; Dablanca-Blanco, A.B.; Varela-Aneiros, I.; Castelo-Baz, P. Effectiveness of different irrigant activation systems on smear layer removal: A scanning electron microscopic study. *J. Clin. Med.* **2022**, *11*, 1003. [CrossRef] [PubMed]
37. Rossi-Fedele, G.; Guastalli, A.R.; Doğramaci, E.J.; Steier, L.; de Figueiredo, J.A.P. Influence of pH changes on chlorine-containing endodontic irrigating solutions. *Int. Endod. J.* **2011**, *44*, 792–799. [CrossRef]

Journal of
Functional Biomaterials

MDPI

Article

Physicochemical and Antibacterial Properties of Conventional and Two Premixed Root Canal Filling Materials in Primary Teeth

Claire El Hachem [1], Jean Claude Abou Chedid [1], Walid Nehme [2], Marc Krikor Kaloustian [3], Nabil Ghosn [4], Hafsa Sahnouni [5], Davide Mancino [5,6,7], Youssef Haikel [5,6,7] and Naji Kharouf [5,6,*]

1 Department of Pediatric Dentistry, Faculty of Dentistry, Saint Joseph University, Beirut 1107 2180, Lebanon
2 Department of Endodontics, Arthur A. Dugoni School of Dentistry, University of the Pacific, 155 5th Street, San Francisco, CA 94103, USA
3 Department of Endodontics, Faculty of Dentistry, Saint Joseph University, Beirut 1107 2180, Lebanon
4 Craniofacial Research Laboratory, Faculty of Dental Medicine, Saint Joseph University, Beirut 1107 2180, Lebanon
5 Department of Biomaterials and Bioengineering, INSERM UMR_S 1121, Biomaterials and Bioengineering, 67000 Strasbourg, France
6 Department of Endodontics, Faculty of Dental Medicine, Strasbourg University, 67000 Strasbourg, France
7 Pôle de Médecine et Chirurgie Bucco-Dentaire, Hôpital Civil, Hôpitaux Universitaire de Strasbourg, 67000 Strasbourg, France
* Correspondence: dentistenajikharouf@gmail.com; Tel.: +33-667522841

Abstract: In this study, some physicochemical and antibacterial properties of three root canal filling materials for primary teeth, Calplus "CP" (Prevest DenPro, Lewes, DE, USA), Bio-C Pulpecto "Bio-CP" (Angelus, Basil, Londrina, Paraná, Brazil), and Zinc Oxide and Eugenol "ZOE" (Prevest DenPro, Lewes, DE, USA) were compared. For each material, the pH, solubility, contact angle, and crystalline microstructure under SEM were evaluated. Their antibacterial activity against *Enterococcus faecalis* was determined through direct tests. The Kruskal–Wallis test was used to analyze the results using a one-way analysis of variance on ranks. All the materials had an alkaline pH at 3, 24, and 72 h, with CalPlus having the highest ($p < 0.05$). Bio-CP was more soluble during the evaluation period (24 h) than ZOE and CalPlus ($p < 0.05$). Bio-CP and ZOE demonstrated the creation of crystallite structures on their surfaces after immersion in PBS at 37 °C, whereas CalPlus showed none. The lowest contact angle was observed for Bio-CP (53 ± 1.5°); contact angles of (86 ± 4°) and (96 ± 1°), respectively, were observed after 10 s of the deposition of the water drop for CalPlus and ZOE. In conclusion, according to this study, there is still a need to develop new filling materials for primary teeth. ZOE, CalPlus and Bio-CP demonstrated different physicochemical and antibacterial properties, but none of the materials had optimal properties and could be considered the most suitable filling material for primary teeth pulpectomy. Bioceramics in their current state are not an alternative. The physicochemical and antibacterial properties still need improvement to fit the intricate anatomy of primary teeth.

Keywords: deciduous tooth; calcium silicate material; calcium hydroxide and iodoform; zinc oxide eugenol; root canal filling

1. Introduction

When faced with an irreversibly infected or necrotic pulp, pulpectomy aims to retain primary teeth in a functional, healthy state until their physiological exfoliation [1]. The ultimate root canal filling material should be antibacterial, and resorbable at the same rate as the primary roots or if extruded. Moreover, it should bond to canal walls without shrinking, and cause no harm to the periapical area or the developing tooth germ [2].

A plethora of materials have been proposed over the years like zinc oxide eugenol alone or with formocresol, iodoform, and camphorated phenol. Pastes containing calcium hydroxide, iodoform, or a mixture of both have also been suggested [3–6].

Zinc oxide eugenol (ZOE) cement, a combination of zinc oxide (powder) and eugenol (liquid) has been, until recently, the most common filling material used [7]. However, it sets into a thick mass that resists resorption, may irritate periapical tissues, and can cause deviation of the permanent tooth bud [8].

Oil-based calcium hydroxide pastes containing iodoform, such as CalPlus (DenPro, Prevest, USA), have also been advocated as filling materials for primary teeth due to their antibacterial and healing properties, and easy handling [9]. When compared to the conventional filling materials (ZOE), this product is premixed and pre-dosed. It does not require any manual mixing procedure [10]. Their main disadvantage is potential external resorption and intracanal resorption, which can ultimately lead to failure [11].

New endodontic types of cement, such as bioceramics, have been gaining in popularity because of their physicochemical and biological properties, including alkaline pH, shrink-free property, chemical stability in the biological environment, biocompatibility, and bioactivity [12]. Bioceramics are composed mainly of calcium silicate materials and are a substitute for the traditionally used calcium hydroxide [13]. These bioactive materials are used in permanent teeth in different applications such as pulpotomy, pulp capping, resorption, perforation repair, and root canal filling [14–17]. They have also been recently advocated in pediatric dentistry [18].

Bio-C Pulpecto (Bio-CP) (Angelus, Basil, Londrina, Paraná, Brazil) is the first resorbable bioceramic root canal filling material for primary teeth. It is composed of titanium oxide, ester glycol salicylate, silicon dioxide, calcium tungstate, toluene sulphonamide, and calcium silicate [19]. To date, there is no comparative study between Bio-CP, ZOE and CalPlus paste regarding their physicochemical and antibacterial properties.

The obturation technique of primary teeth relies solely on the filling material, hence the importance of the antibacterial activity of the filling material. One of the most prevalent species resistant to mechanical preparation and irrigation protocols identified in human primary teeth is *Enterococcus faecalis* [20–22]. Moreover, when opting for a filling material, there should be a complete understanding of the physical and chemical behavior of the material in order to choose the most appropriate for every clinical situation.

The aim of this study was to evaluate, in vitro, some physicochemical properties and the antibacterial activity of ZOE, CalPlus, and Bio-C Pulpecto as resorbable filling materials to determine the most appropriate filling material for primary teeth pulpectomy. The null hypothesis was that there would be no significant difference in the physicochemical and/or antibacterial characteristics between the three tested materials.

2. Materials and Methods

2.1. Materials

Calplus "CP" (DenPro, Prevest, USA), Bio-C Pulpecto "Bio-CP" (Angelus, Basil, Londrina, Paraná, Brazil), and Zinc Oxide and Eugenol "ZOE" (DenPro, Prevest, USA) were used in the present study following the manufacturer's instructions (Table 1). All specimens were conserved in a dark container for 48 h at 37 °C to accomplish a complete setting time [23].

Table 1. Manipulation and manufacturer and of tested materials.

Materials	Manufacturer	Lot	Mixing
Calplus (CP)	Prevest DenPro, Lewes, DE, USA	3121906	Premixed
Bio-C Pulpecto (Bio-CP)	Angelus, Basil, Londrina, Paraná, Brazil	51152	Premixed
Zinc Oxide and Eugenol (ZOE)	Prevest DenPro, Lewes, DE, USA	1561912	1.4 g zinc oxide to 0.4 mL eugenol

2.2. pH Measurements of the Aqueous Solution in Contact with the Cement

For each group, five samples were prepared using Teflon molds (3 mm in diameter and 3.8 in height). Each sample was put in contact with 10 mL of distilled water at 37 °C. A pH meter "CyberScan pH 510" (Thermo Scientific, Waltham, MA, USA) was used to measure the pH of water in contact with each sample at 3, 24, and 72 h. Distilled water was flushed over the pH meter electrode to eliminate contamination from the previous solution.

2.3. Solubility

Three samples (20 mm in diameter and 2 mm in height) of each group were prepared and analyzed following a previous study [24]. Using a digital system (accuracy ± 0.0001 g), the samples were weighed three times before the 24 h of the immersion period in 50 mL of distilled water at 37 °C. After the aging period, the samples were taken out from distilled water and then dried at 37 °C for 24 h. Finally, the samples were weighed again three times and averaged to obtain the final weight. For each material, the solubility percentage was attained from the difference in mass between the final and the initial weight.

2.4. Scanning Electron Microscope (SEM) of Crystallites Creation

Twelve samples for each material were prepared as described in Section 2.2. From each group, three samples were stored in hermetic boxes and kept in dry conditions. The remaining samples from each group were put in 10 mL of phosphate-buffered saline (PBS10×, Dominique Dutscher, Bernolsheim, France) at 37 °C. Three periods were assigned (24, 72, and 168 h) to investigate the morphological changes of the cement surfaces using SEM analysis. After each period, distilled water was poured over the samples for 5 min. These were mounted on SEM stubs and sputter-coated with gold–palladium (20/80), then analyzed using an SEM (FEI Company, Eindhoven, The Netherlands, 10 kV) at a magnification of ×5000 with a working distance of 10 mm [25].

2.5. Water Sorption Tests

From each group, three samples were prepared using Teflon molds (10 mm in diameter and 2 mm in height). After the setting time, the samples were kept dry in the fume hood overnight. To evaluate the sorption time of a 5 µL drop of distilled water into the cement surface, a contact angle device (Biolin Scientific, Espoo, Finland) was used [14]. Using a horizontal camera, a movie was recorded to track the profile of the water drop and its absorption time.

2.6. Antimicrobial Activity

Enterococcus faecalis (*E. faecalis*, ATCC 29212) was cultured in a Brain Heart Infusion medium (BHI) (Darmstadt, Germany). The turbidity of the bacterial medium containing *E. faecalis* was adapted at $OD_{600(nm)} = 0.3$. To evaluate the antibacterial activity of these against *E. faecalis*, a direct contact test (DCT) was used. Each sample was placed in a well (24-well culture plate) (in triplicate). One mL of the bacterial medium was added to each well and incubated in anaerobic conditions at 37 °C for 24 h under constant stirring at 450 rpm [26]. The bacterial medium without the tested filling materials was used as a control group. After 24 h, on each specimen from each group, 10-fold serial dilutions up to 10^6 in BHI were performed. A volume of 100 µL of each diluted medium was added onto a BHI agar plate, then homogeneously spread and incubated at 37 °C for 24 h. Manual counting was used to measure the *E. faecalis* concentration by counting the colonies on the plate, and their CFU/mL (colony forming units/mL) was determined.

2.7. Statistical Analysis

The SigmaPlot release 11.2 (Systat Software, Inc., San Jose, CA, USA) was used to analyze the data. The normality of the data was verified with the Shapiro–Wilk test. To determine whether significant differences existed in the antibacterial activity, pH values, solubility, and angle contact, the Kruskal–Wallis test (one-way analysis of variance on ranks) including multiple comparison procedures (Tukey Test) was used. A statistical significance level was set at $\alpha = 0.05$.

3. Results

3.1. pH Analysis

The pH of the solution in contact with the different types of cement was described for 72 h in Figure 1. All three types of cement demonstrated an alkaline pH for the solution in contact at 3, 24, and 72 h. CalPlus demonstrated the highest alkaline pH during 72 h compared to the two other types of cement. At 3 h, CalPlus has higher alkaline pH than ZOE ($p < 0.05$). At 24 h, a significant difference was found between the three tested materials ($p < 0.05$). Finally, at 72 h, ZOE has a higher pH than Bio-CP with no difference compared to CalPlus.

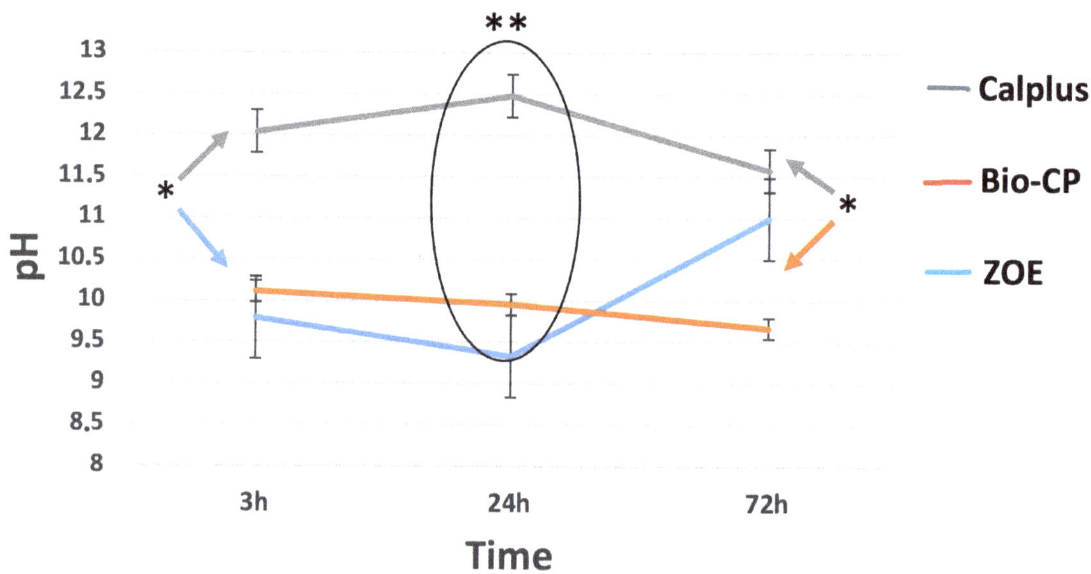

Figure 1. pH changes of the water in contact with the different tested types of cement after 3, 24, and 72 of immersion in distilled water at 37 °C. Calplus (CP), Bio-C Pulpecto (Bio-CP), and Zinc Oxide and Eugenol (ZOE). (* $p = 0.002$; ** $p \leq 0.001$).

3.2. Solubility

For each tested material, the mean and standard deviation of solubility (wt.%) values are presented in Figure 2. Bio-CP was more soluble during the evaluation period (24 h) than ZOE and CalPlus ($p < 0.05$).

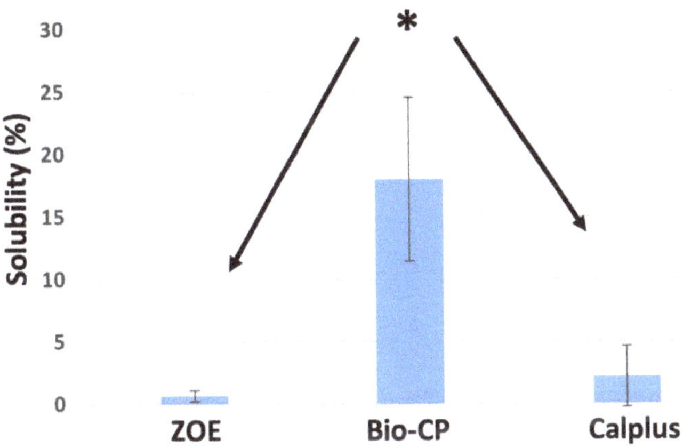

Figure 2. Solubility percentages of the different types of cement after immersion in 50 mL of distilled water at 37 °C for 24 h. * $p < 0.05$.

3.3. Scanning Electron Microscope (SEM)

The crystalline structures of the three types of cement are shown in Figure 3. Bio-CP and ZOE demonstrated the creation of crystallite structures on their surfaces after immersion in PBS at 37 °C. Some zones of ZOE surface demonstrated cubical crystalline structures, whilst Bio-CP showed an urchin-like crystallite structure. CalPlus did not show any crystallite creation.

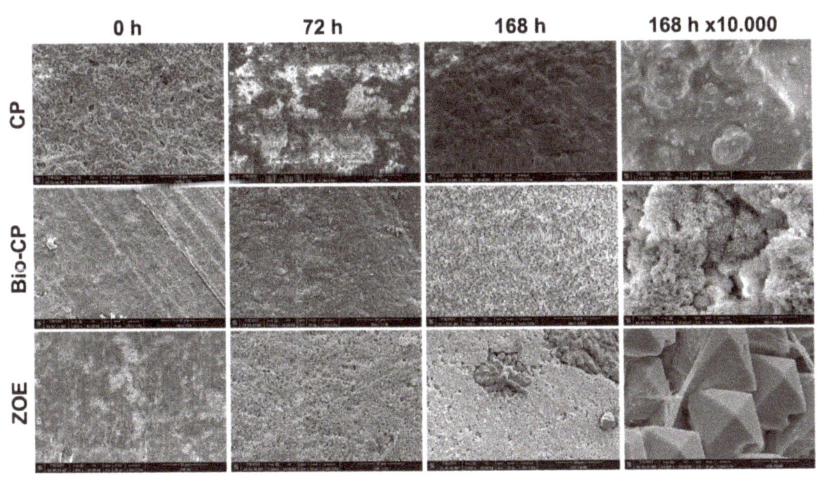

Figure 3. Scanning electron microscope images (×1000 and ×10,000 magnifications) demonstrate the morphological changes of each material at different time points of immersion in phosphate-buffered solution at 37 °C.

3.4. Water Sorption Tests

Bio-CP demonstrated the highest hydrophilicity for 5 uL of a drop of distilled water compared to ZOE and CalPlus. The lowest contact angle was observed for Bio-CP (53 ± 1.5°). Contact angles of (86 ± 4°) and (96 ± 1°), respectively were observed after 10 s of deposition of the water drop for CalPlus and ZOE (Figure 4, Table 2).

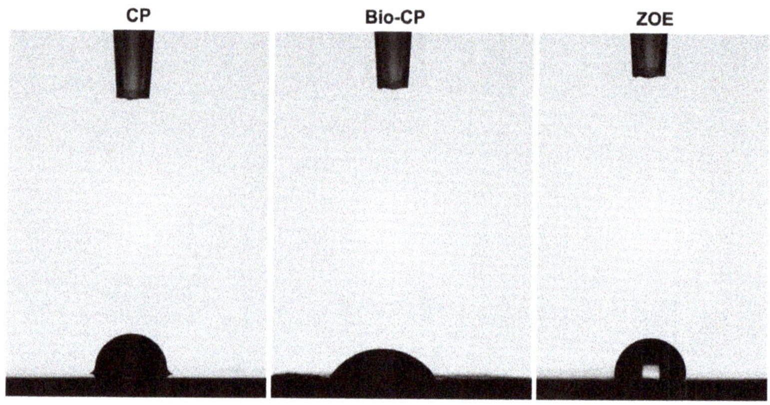

Figure 4. Contact angles after 5 µL of water dropped on the different cement surfaces after 10 s of deposition. Calplus (CP), Bio-C Pulpecto (Bio-CP), and Zinc Oxide and Eugenol (ZOE).

Table 2. Contact angles of 5 µL of distilled water on the different material surfaces after 10 s of deposition. Statistical significance ($p < 0.05$) is indicated with superscript letters a, b and c. Calplus (CP), Bio-C Pulpecto (Bio-CP), and Zinc Oxide and Eugenol (ZOE).

Test\Materials	CalPlus	Bio-CP	ZOE	Statistical Significance
Contact angle (°)	86 ± 4 [a]	53 ± 1.5 [b]	96 ± 1 [c]	$p < 0.05$

3.5. Antimicrobial Activity

Significant bacterial growth was observed after 24 h between all the groups ($p < 0.05$). ZOE is the most effective, killing 100% of bacteria, followed by Bio-CP, which eliminates more than 75%, and finally CalPlus (more than 50% of bacteria) ($p < 0.05$) (Figure 5). In addition, all three types of cement have antibacterial activity against *E. faecalis* and the best-performing cement was ZOE.

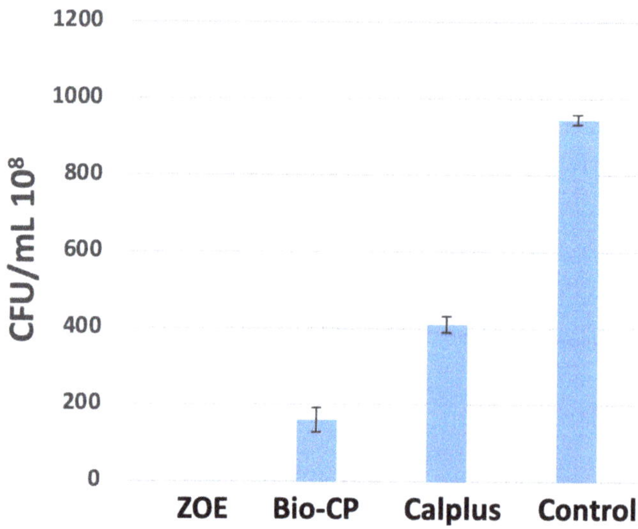

Figure 5. The number of colonies forming units/mL of *Enterococcus faecalis* for the medium in contact with Calplus "CP", Bio-C Pulpecto "Bio-CP", Zinc Oxide, and Eugenol "ZOE" and the control group "C" (bacterial medium) after 24 h at 37 °C in anaerobic conditions.

4. Discussion

The obturation of primary teeth relies completely on the filling material [27]. ZOE has been, until 2008, the only material explicitly recommended by the American Academy of Pediatric Dentistry (AAPD) [28]. However, it cannot be regarded as an ideal root canal filling due to its resorption rate and negative effects when extruded [3]. Calcium hydroxide and iodoform pastes were recently recommended and showed a high success rate in randomized clinical trials [29]. Nonetheless, in a recent systematic review and meta-analysis, an unclear or high risk of bias was proven in most of the studies and the overall certainty of the evidence ranged from low to very low, denoting that there is still no conclusion to be drawn as to the best pulpectomy material [9].

A resorbable Bioceramic paste, like Bio-CP, could be a promising alternative if the material possessed physicochemical and antibacterial properties suitable for the filling of the intricate root canal system of primary teeth [30]. This was the first in vitro study in the pediatric endodontics literature to focus on the pH, solubility, contact angle, crystallographic changes using SEM images after immersion in PBS, and antibacterial activity of three filling materials for primary teeth. The results of this study showed significant differences between ZOE, CalPlus, and Bio-CP for all evaluated criteria; thus, the null hypothesis was rejected.

The pH analysis indicated an alkaline pH for all tested materials in contact with distilled water at 24 h. An alkaline pH plays a role in the deposition of mineralized tissue and neutralizing the lactic acid from the osteoclast, and promoting healing [31]. Higher pH values were observed for CalPlus compared to ZOE and Bio-CP ($p < 0.05$), which can be attributed to its dissociation into Ca^2+ and OH^- ions and has already been validated in many previous studies [32,33]. However, the main disadvantage of this type of material is intracanal resorption and accelerated external root resorption reported in several clinical studies [27].

A set endodontic sealer should present solubility of less than 3% according to ISO standards [23]. In our study, Bio-CP exhibited very high solubility exceeding 3%, demonstrating a high level of Ca^{2+} and bioactivity. However, high solubility also means that the material dissolves quickly, creating gaps in the canals and generating reinfection [34]. This was also observed in other studies about bioceramic sealers [35] and could be explained by the presence of fine hydrophilic particles in the composition of these sealers, which generate an increase in surface area that may increase sealer solubility when in contact with water/moisture, [36]. Therefore, it is essential to determine clinically the rate at which Bio-CP dissolves, since there is no Gutta Percha to complement the sealer in the root canal treatment of primary teeth. Moreover, the lifespan of a root canal treatment varies greatly depending on the child's age and physiologic stage of the roots. In the only other study about Bio-CP, the authors highlighted that Bio-CP exhibited sufficient physicochemical properties, disclosed cytocompatibility, and indicated the potential to stimulate mineralization, but lacked clinical support [19].

Using SEM analysis, the microstructural crystalline formation during the initial setting time in 95% humidity and after 3 and 7 days in water was observed. CalPlus did not demonstrate a clear crystallite formation on its surface (Figure 3). This could be because this material is a non-setting paste and the acquisition of such an image was not possible. This result could not be compared to other studies since, to the best of our knowledge, there has not been a publication about the microstructure of calcium hydroxide iodoform-based materials.

The formation of a crystalline structure stipulates that remineralization can occur [23]. In contrast, ZOE had some zones with cubical structures, whereas Bio-CP had an urchin-like structure on its surface. This aspect was extensively described in many studies and was attributed to the extraction of hydroxyapatite from natural resources [37,38].

Contact angle measurements are a reliable tool to better understand the interactions between solids and liquids [39]. These interactions play a major role in explaining not only material wettability, but also wetting, spreading, and adsorption of liquids [39]. In this study, Bio-CP had the lowest contact angle (53 ± 1.5°), demonstrating a good wetting ability, and the capacity to spread faster on substrates such as dentin [40]. CalPlus and ZOE both exhibited poor surface wetting with high contact angles of (86 ± 4°) and (96 ± 1°), respectively. Tummala et al. also found a high contact angle for ZOE and credited this to the increased viscosity of the sealer [41].

The antibacterial potential of each material against *E. faecalis* was evaluated using a direct contact test per previous studies [42]. Anaerobic bacteria were found scattered within the whole root canal system of primary teeth (accessory canals, dentinal tubules, secondary canals, apical foramen) as well as on the physiological resorptive zones, hence the crucial role of the antibacterial property of the filling paste [43]. This species of bacteria may persist even after biomechanical preparation and use of intracanal irrigants [44]. According to this study, ZOE, CalPlus, and Bio-CP possess antibacterial activity compared to the bacterial medium ($p < 0.05$). ZOE is the most efficient ($p < 0.05$), corroborating the results of previous studies [45]. Eugenol-based root canal filling materials possess antibacterial activity due to the action of eugenol, triggering protein denaturation and rendering the microorganisms non-functional [46]. However, in some studies, ZOE demonstrated the smallest zones of bacterial growth inhibition against *E. faecalis* [43], and higher antibacterial activity for iodoform pastes was observed [47].

CalPlus iodoform paste did not set; therefore, compressive tests were not carried out. Furthermore, the filling ability, flow, radio-opacity and biological properties of each material were omitted [23], and they are important factors to consider. Further in-vitro studies should be conducted to better understand the physical and chemical properties of filling materials for primary teeth. Ageing of Bio-CP in simulated body fluid could provide interesting data about the resorption rate compared to that of the roots of primary teeth [48].

The results obtained in this study are encouraging for the use of resorbable bioceramics in pediatric endodontics. Bio-CP contains calcium silicate and can be classified as a bioactive material. It is capable of forming hydroxyapatite or carbonated apatite on its surface and inducing osteogenesis [48].

However, the results of this study are not conclusive and there is still a need to conduct more well-designed studies to better understand the credentials of an ideal filling material. For the Bioceramics, there is still a need to improve their properties, mainly the excessive solubility, and to increase their antibacterial effect. Moreover, due to the continuous root resorption occurring on primary teeth, there is a need for randomized long-term clinical trials to assess the clinical behavior of this type of material.

5. Conclusions

Within the limitations of this in-vitro study, it could be concluded that there is still a need to develop new filling materials for the root canal treatment of primary teeth. Given the results of this study, ZOE, CalPlus and Bio-CP demonstrated different physicochemical and antibacterial properties, but none of the materials had optimal properties and could be considered the most suitable filling material for primary teeth pulpectomy. The properties of bioceramics such as bioactivity, solubility in fluids, and adhesiveness would provide a crucial step in increasing the success rate of root canal treatment on primary teeth and developing more performant materials. More research should be conducted to optimize clinical protocols in pediatric endodontics.

Author Contributions: Conceptualization, C.E.H. and N.K.; methodology, N.G. and H.S.; validation, D.M. and Y.H.; writing—original draft preparation, C.E.H. and M.K.K.; writing—review and editing, J.C.A.C., N.K. and W.N.; supervision, N.K. All authors have read and agreed to the published version of the manuscript.

Funding: This research received no external funding.

Institutional Review Board Statement: Not applicable.

Informed Consent Statement: Not applicable.

Data Availability Statement: Not applicable.

Acknowledgments: We acknowledge INSERM UMR_S 1121 laboratory for providing the experimental setups.

Conflicts of Interest: The authors declare no conflict of interest.

References

1. Coll, J.A.; Vargas, K.; Marghalani, A.A.; Chen, C.-Y.; AlShamali, S.; Dhar, V.; Crystal, Y.O. A Systematic Review and Meta-Analysis of Nonvital Pulp Therapy for Primary Teeth. *Pediatr. Dent.* **2020**, *42*, 256–461. [PubMed]
2. Shindova, M. Root canal filling materials in primary teeth—Review. *Folia Med.* **2021**, *63*, 657–662. [CrossRef]
3. Huang, T.-H.; Hung, C.-J.; Chen, Y.-J.; Chien, H.-C.; Kao, C.-T. Cytologic effects of primary tooth endodontic filling materials. *J. Dent. Sci.* **2009**, *4*, 18–24. [CrossRef]
4. Subramaniam, P.; Gilhotra, K. Endoflas, Zinc Oxide Eugenol and Metapex as Root Canal Filling Materials in Primary Molars—A Comparative Clinical Study. *J. Clin. Pediatr. Dent.* **2011**, *35*, 365–370. [CrossRef] [PubMed]
5. Reddy, S.; Ramakrishna, Y. Evaluation of antimicrobial efficacy of various root canal filling materials used in primary teeth: A microbiological study. *J. Clin. Pediatr. Dent.* **2007**, *31*, 193–198. [CrossRef]
6. Pilownic, K.J.; Gomes, A.P.N.; Wang, Z.J.; Almeida, L.H.S.; Romano, A.R.; Shen, Y.; Felix, A.d.O.C.; Haapasalo, M.; Pappen, F.G. Physicochemical and Biological Evaluation of Endodontic Filling Materials for Primary Teeth. *Braz. Dent. J.* **2017**, *28*, 578–586. [CrossRef]
7. Barcelos, R.; Santos, M.P.A.; Primo, L.G.; Luiz, R.R.; Maia, L.C. ZOE paste pulpectomies outcome in primary teeth: A systematic review. *J. Clin. Pediatr. Dent.* **2011**, *35*, 241–248. [CrossRef]
8. Trairatvorakul, C.; Chunlasikaiwan, S. Success of Pulpectomy With Zinc Oxide-Eugenol Vs Calcium Hydroxide/Iodoform Paste in Primary Molars: A Clinical Study. *Pediatr. Dent.* **2008**, *30*, 303–308.
9. Silva Junior, M.F.; Wambier, L.M.; Gevert, M.V.; Chibinski, A.C.R. Effectiveness of iodoform-based filling materials in root canal treatment of deciduous teeth: A systematic review and meta-analysis. *Biomater. Investig. Dent.* **2022**, *9*, 52–74. [CrossRef]
10. Eid, A.; Davide, M.; Joudi, F.; Rekab, M.; Layous, K.; Hamadah, O.; Haikel, Y.; Kharouf, N. Comparative Evaluation Of The Apical Sealing Ability Of BioRoot And Ahplus Sealers: An In Vitro Study. *Int. J. Dent. Oral Sci.* **2021**, *8*, 2309–2313. [CrossRef]
11. Moskovitz, M.; Tickotsky, N.; Ashkar, H.; Holan, G. Degree of root resorption after root canal treatment with iodoform-containing filling material in primary molars. *Quintessence Int. Berl. Ger. 1985* **2012**, *43*, 361–368.
12. Marciano, M.A.; Duarte, M.A.H.; Camilleri, J. Calcium silicate-based sealers: Assessment of physicochemical properties, porosity and hydration. *Dent. Mater.* **2016**, *32*, e30–e40. [CrossRef] [PubMed]
13. Hardan, L.; Mancino, D.; Bourgi, R.; Alvarado-Orozco, A.; Rodríguez-Vilchis, L.E.; Flores-Ledesma, A.; Cuevas-Suárez, C.E.; Lukomska-Szymanska, M.; Eid, A.; Danhache, M.-L.; et al. Bond Strength of Adhesive Systems to Calcium Silicate-Based Materials: A Systematic Review and Meta-Analysis of In Vitro Studies. *Gels Basel Switz.* **2022**, *8*, 311. [CrossRef]
14. Kharouf, N.; Zghal, J.; Addiego, F.; Gabelout, M.; Jmal, H.; Haikel, Y.; Bahlouli, N.; Ball, V. Tannic acid speeds up the setting of mineral trioxide aggregate cements and improves its surface and bulk properties. *J. Colloid Interface Sci.* **2021**, *589*, 318–326. [CrossRef]
15. Kharouf, N.; Haïkel, Y.; Mancino, D. Unusual Maxillary First Molars with C-Shaped Morphology on the Same Patient: Variation in Root Canal Anatomy. *Case Rep. Dent.* **2019**, *2019*, 1857289. [CrossRef] [PubMed]
16. Eid, A.; Mancino, D.; Rekab, M.S.; Haikel, Y.; Kharouf, N. Effectiveness of Three Agents in Pulpotomy Treatment of Permanent Molars with Incomplete Root Development: A Randomized Controlled Trial. *Healthc. Basel Switz.* **2022**, *10*, 431. [CrossRef] [PubMed]
17. Kassab, P.; El Hachem, C.; Habib, M.; Nehme, W.; Zogheib, C.; Tonini, R.; Kaloustian, M.K. The Pushout Bond Strength of Three Calcium Silicate-based Materials in Furcal Perforation Repair and the Effect of a Novel Irrigation Solution: A Comparative In Vitro Study. *J. Contemp. Dent. Pract.* **2022**, *23*, 289–294.
18. Stringhini Junior, E.; Dos Santos, M.G.C.; Oliveira, L.B.; Mercadé, M. MTA and biodentine for primary teeth pulpotomy: A systematic review and meta-analysis of clinical trials. *Clin. Oral Investig.* **2019**, *23*, 1967–1976. [CrossRef] [PubMed]
19. Ochoa Rodríguez, V.M.; Tanomaru-Filho, M.; Rodrigues, E.M.; Bugança, E.d.O.; Guerreiro-Tanomaru, J.M.; Faria, G. Physico-chemical properties and effect of bioceramic root canal filling for primary teeth on osteoblast biology. *J. Appl. Oral Sci.* **2021**, *29*, e20200870. [CrossRef]
20. Cancio, V.; Carvalho Ferreira, D.D.; Cavalcante, F.S.; Rosado, A.S.; Teixeira, L.M.; Braga Oliveira, Q.; Barcelos, R.; Gleiser, R.; Santos, H.F.; dos Santos, K.R.N.; et al. Can the Enterococcus faecalis identified in the root canals of primary teeth be a cause of failure of endodontic treatment? *Acta Odontol. Scand.* **2017**, *75*, 423–428. [CrossRef]
21. Forghani, M.; Afshari, E.; Parisay, I.; Garajian, R. Effect of a passive sonic irrigation system on elimination of Enterococcus faecalis from root canal systems of primary teeth, using different concentrations of sodium hypochlorite: An in vitro evaluation. *J. Dent. Res. Dent. Clin. Dent. Prospect.* **2017**, *11*, 177–182. [CrossRef] [PubMed]

22. Cogulu, D.; Uzel, A.; Oncag, O.; Eronat, C. PCR-based identification of selected pathogens associated with endodontic infections in deciduous and permanent teeth. *Oral Surg. Oral Med. Oral Pathol. Oral Radiol. Endodontol.* **2008**, *106*, 443–449. [CrossRef] [PubMed]
23. Kharouf, N.; Arntz, Y.; Eid, A.; Zghal, J.; Sauro, S.; Haikel, Y.; Mancino, D. Physicochemical and Antibacterial Properties of Novel, Premixed Calcium Silicate-Based Sealer Compared to Powder–Liquid Bioceramic Sealer. *J. Clin. Med.* **2020**, *9*, 3096. [CrossRef] [PubMed]
24. Kharouf, N.; Mancino, D.; Zghal, J.; Helle, S.; Jmal, H.; Lenertz, M.; Viart, N.; Bahlouli, N.; Meyer, F.; Haikel, Y.; et al. Dual role of tannic acid and pyrogallol incorporated in plaster of Paris: Morphology modification and release for antimicrobial properties. *Mater. Sci. Eng. C Mater. Biol. Appl.* **2021**, *127*, 112209. [CrossRef] [PubMed]
25. Kharouf, N.; Sauro, S.; Jmal, H.; Eid, A.; Karrout, M.; Bahlouli, N.; Haikel, Y.; Mancino, D. Does Multi-Fiber-Reinforced Composite-Post Influence the Filling Ability and the Bond Strength in Root Canal? *Bioeng. Basel Switz.* **2021**, *8*, 195. [CrossRef]
26. Kharouf, N.; Sauro, S.; Hardan, L.; Fawzi, A.; Suhanda, I.E.; Zghal, J.; Addiego, F.; Affolter-Zbaraszczuk, C.; Arntz, Y.; Ball, V.; et al. Impacts of Resveratrol and Pyrogallol on Physicochemical, Mechanical and Biological Properties of Epoxy-Resin Sealers. *Bioeng. Basel Switz.* **2022**, *9*, 85. [CrossRef]
27. Najjar, R.S.; Alamoudi, N.M.; El-Housseiny, A.A.; Al Tuwirqi, A.A.; Sabbagh, H.J. A comparison of calcium hydroxide/iodoform paste and zinc oxide eugenol as root filling materials for pulpectomy in primary teeth: A systematic review and meta-analysis. *Clin. Exp. Dent. Res.* **2019**, *5*, 294–310. [CrossRef]
28. American Academy of Pediatric Dentistry Clinical Affairs Committee—Pulp Therapy Subcommittee; American Academy of Pediatric Dentistry Council on Clinical Affairs. Guideline on pulp therapy for primary and young permanent teeth. *Pediatr. Dent.* **2005**, *27*, 130–134.
29. Al-Attiya, H.; Schmoeckel, J.; Mourad, M.S.; Splieth, C.H. One year clinical success of pulpectomy in primary molars with iodoform-calcium hydroxide paste. *Quintessence Int. Berl. Ger. 1985* **2021**, *52*, 528–537. [CrossRef]
30. El Hachem, C.; Kaloustian, M.K.; Nehme, W.; Ghosn, N.; Abou Chedid, J.C. Three-dimensional modeling and measurements of root canal anatomy in second primary mandibular molars: A case series micro CT study. *Eur. Arch. Paediatr. Dent. Off. J. Eur. Acad. Paediatr. Dent.* **2019**, *20*, 457–465. [CrossRef]
31. Paz, L.C.d. Redefining the Persistent Infection in Root Canals: Possible Role of Biofilm Communities. *J. Endod.* **2007**, *33*, 652–662. [CrossRef] [PubMed]
32. Tronstad, L.; Andreasen, J.O.; Hasselgren, G.; Kristerson, L.; Riis, I. pH changes in dental tissues after root canal filling with calcium hydroxide. *J. Endod.* **1981**, *7*, 17–21. [CrossRef]
33. Foreman, P.C.; Barnes, I.E. A review of calcium hydroxide. *Int. Endod. J.* **1990**, *23*, 283–297. [CrossRef] [PubMed]
34. Torres, F.F.E.; Zordan-Bronzel, C.L.; Guerreiro-Tanomaru, J.M.; Chávez-Andrade, G.M.; Pinto, J.C.; Tanomaru-Filho, M. Effect of immersion in distilled water or phosphate-buffered saline on the solubility, volumetric change and presence of voids within new calcium silicate-based root canal sealers. *Int. Endod. J.* **2020**, *53*, 385–391. [CrossRef]
35. Poggio, C.; Dagna, A.; Ceci, M.; Meravini, M.-V.; Colombo, M.; Pietrocola, G. Solubility and pH of bioceramic root canal sealers: A comparative study. *J. Clin. Exp. Dent.* **2017**, *9*, e1189–e1194. [CrossRef] [PubMed]
36. Mann, A.; Zeng, Y.; Kirkpatrick, T.; van der Hoeven, R.; Silva, R.; Letra, A.; Chaves de Souza, L. Evaluation of the Physicochemical and Biological Properties of EndoSequence BC Sealer HiFlow. *J. Endod.* **2022**, *48*, 123–131. [CrossRef]
37. Eliaz, N.; Metoki, N. Calcium Phosphate Bioceramics: A Review of Their History, Structure, Properties, Coating Technologies and Biomedical Applications. *Materials* **2017**, *10*, 334. [CrossRef]
38. Akram, M.; Ahmed, R.; Shakir, I.; Ibrahim, W.A.W.; Hussain, R. Extracting hydroxyapatite and its precursors from natural resources. *J. Mater. Sci.* **2014**, *49*, 1461–1475. [CrossRef]
39. Kontakiotis, E.G.; Tzanetakis, G.N.; Loizides, A.L. A comparative study of contact angles of four different root canal sealers. *J. Endod.* **2007**, *33*, 299–302. [CrossRef]
40. Prado, M.; de Assis, D.F.; Gomes, B.P.F.A.; Simão, R.A. Effect of Disinfectant Solutions on the Surface Free Energy and Wettability of Filling Material. *J. Endod.* **2011**, *37*, 980–982. [CrossRef]
41. Tummala, M.; Chandrasekhar, V.; Rashmi, A.S.; Kundabala, M.; Ballal, V. Assessment of the wetting behavior of three different root canal sealers on root canal dentin. *J. Conserv. Dent. JCD* **2012**, *15*, 109–112. [CrossRef] [PubMed]
42. Zhang, H.; Shen, Y.; Ruse, N.D.; Haapasalo, M. Antibacterial Activity of Endodontic Sealers by Modified Direct Contact Test Against Enterococcus faecalis. *J. Endod.* **2009**, *35*, 1051–1055. [CrossRef] [PubMed]
43. Queiroz, A.M.D.; Nelson-Filho, P.; Silva, L.A.B.D.; Assed, S.; Silva, R.A.B.D.; Ito, I.Y. Antibacterial activity of root canal filling materials for primary teeth: Zinc oxide and eugenol cement, Calen paste thickened with zinc oxide, Sealapex and EndoREZ. *Braz. Dent. J.* **2009**, *20*, 290–296. [CrossRef] [PubMed]
44. Moradi, F.; Haghgoo, R. Evaluation of Antimicrobial Efficacy of Nanosilver Solution, Sodium Hypochlorite and Normal Saline in Root Canal Irrigation of Primary Teeth. *Contemp. Clin. Dent.* **2018**, *9*, S227–S232. [CrossRef] [PubMed]
45. Pimenta, H.C.; Borges, Á.H.; Bandeca, M.C.; Neves, A.T.S.; Fontes, R.G.; da Silva, P.V.; Aranha, A.M.F. Antimicrobial Activity of Filling Materials Used in Primary Teeth Pulpotomy. *J. Int. Oral Health* **2015**, *7*, 54–57.
46. Verma, R.; Sharma, D.; Pathak, A. Antibacterial Efficacy of Pastes Against E Faecalis in Primary Root Dentin: A Confocal Microscope Study. *J. Clin. Pediatr. Dent.* **2015**, *39*, 247–254. [CrossRef]

47. Estrela, C.; Estrela, C.R.D.A.; Hollanda, A.C.B.; Decurcio, D.D.A.; Pécora, J.D. Influence of iodoform on antimicrobial potential of calcium hydroxide. *J. Appl. Oral Sci.* **2006**, *14*, 33–37. [CrossRef]
48. Elsayed, M.A.; Hassanien, E.E.; Elgendy, A.A.E. Ageing of TotalFill BC Sealer and MTA Fillapex in Simulated Body Fluid. *Eur. Endod. J.* **2021**, *6*, 183–188. [CrossRef]

Journal of
Functional Biomaterials

MDPI

Article

Evaluation of Effects of Various Irrigating Solutions on Chemical Structure of Root Canal Dentin Using FTIR, SEM, and EDS: An In Vitro Study

Indu Padmakumar [1], Dharam Hinduja [1], Abdul Mujeeb [1,*], Raghu Kachenahalli Narasimhaiah [1], Ashwini Kumar Saraswathi [1], Mubashir Baig Mirza [2], Ali Robaian [2], Syed Nahid Basheer [3], Mohmed Isaqali Karobari [4,5,*] and Giuseppe Alessandro Scardina [6,*]

1 Department of Conservative Dentistry and Endodontics, SJM Dental College and Hospital, Chitradurga 577501, Karnataka, India
2 Conservative Dental Science Department, College of Dentistry, Prince Sattam bin Abdulaziz University, AlKharj 11942, Saudi Arabia
3 Department of Restorative Dental Sciences, College of Dentistry, Jazan University, Jazan 45142, Saudi Arabia
4 Department of Conservative Dentistry & Endodontics, Saveetha Dental College & Hospitals, Saveetha Institute of Medical and Technical Sciences University, Chennai 600077, Tamil Nadu, India
5 Department of Restorative Dentistry & Endodontics, Faculty of Dentistry, University of Puthisastra, Phnom Penh 12211, Cambodia
6 Department of Surgical, Oncological and Stomatological Disciplines, University of Palermo, 90133 Palermo, Italy
* Correspondence: drabdulmujeebmds@gmail.com (A.M.); dr.isaq@gmail.com (M.I.K.); alessandro.scardina@unipa.it (G.A.S.)

Abstract: Background: Sequential chemical application for irrigating a root canal during chemomechanical debridement can affect the dentin microstructure. Understanding the effects of various irrigants on chemical properties of dentin can elucidate their effects on physical properties and thereby explain the higher incidence of structural failure in endodontically treated teeth. This in vitro research aimed to compare and evaluate the effects of three different irrigating solutions on the chemical structure of root canal dentin in extracted human teeth. **Methods:** Forty-eight extracted single-rooted mandibular premolar teeth were sectioned at the cemento–enamel junction by a diamond disc and were then randomly assigned to four groups of twelve samples each. The groups were irrigated using 5.25% NaOCl, ozonated olive oil, silver citrate, or distilled water. Dentin sections measuring 1.5 mm were obtained from the root portion and each section and were analyzed using Fourier transform infrared spectroscopy (FTIR), scanning electron microscopy (SEM), and electron-dispersive spectroscopy (EDS). FTIR and EDS values are reported as means ± standard deviations. Data were analyzed using an ANOVA and a post hoc Bonferroni test ($p < 0.05$). **Results:** A comparison of the FTIR and EDS values among the groups using ANOVA revealed statistically significant differences in the organic and inorganic peak values among the groups. An intergroup comparison between NaOCl with silver citrate and ozonated olive oil revealed significant reductions in the carbonate and phosphate peak values in the NaOCl group ($p < 0.05$). The EDS values tabulated for the carbon, oxygen, phosphorous, and calcium peak levels showed significant differences between the groups using an ANOVA. An SEM analysis was conducted under 1500× magnification, which revealed smear layer removal in the silver citrate group. **Conclusions**: The silver citrate solution and the ozonated olive oil caused less changes in the organic and mineral contents of dentin than sodium hypochlorite.

Keywords: endodontics; irrigants; root dentin; sodium hypochlorite; silver citrate; ozonated olive oil; chemical properties

1. Introduction

Chemomechanical debridement plays an important role in the success of root canal treatments. This is performed by utilizing appropriate instruments along with effective irrigating solutions, followed by sealing with suitable materials [1]. A successful root canal treatment requires irrigation, as it fulfils several mechanical, chemical, and microbiological functions, including healing the periapical tissues. Irrigation thus plays a central role in endodontic treatment, though there is no single irrigant that efficiently satisfies all the required functions of an ideal irrigating solution [2]. The tissue dissolution capacity and the antimicrobial effect are two vital features of endodontic irrigants that enable them to play an integral part in chemomechanical root canal preparation. While enhancing the elimination of microbiota and facilitating the elimination of necrotic tissue and dentin debris from the root canal system, an ideal irrigant should also be nonirritant to the surrounding root and should not debilitate the tooth structure by causing excessive wear of minerals from the dentin [2,3].

Currently, a wide array of endodontic irrigants is available on the market, although sodium hypochlorite continues to be one of the most prominent and widely used endodontic irrigants. It is available in various concentrations ranging from 0.5 to 5.25% [4]. Sodium hypochlorite has been proven to be efficacious in dissolving organic compositions and pulp remnants, and it is the only irrigating solution that has the capability to dissolve vital and nonvital organic tissues [3]. However, it causes considerable destruction to the collagen of surface dentin within a relatively short period of time, which impairs the flexural and elastic strength of the dentin. Additionally, at the end of the chemomechanical preparation, irrigation with hypochlorite causes strong erosion of the surface dentin of the canal wall [5]. Hence, if it is not used judiciously, it could jeopardize the longevity of the root canal treatment. Due to increasing safety concerns, newer irrigants are being studied for the potential replacement of sodium hypochlorite.

In recent studies, a novel endodontic irrigating solution containing silver citrate was developed using electrolytically generated silver ions (0.003%) in citric acid (4.846%) and was tested as an innovative biomaterial for disinfecting and cleaning root canals. The patented aqueous disinfectant is a powerful antimicrobial agent that is produced by an electrochemical process using silver and citric acid that produce a stabilized silver ion complex, which develops a molecular complex, $AgC_6H_7O_7$, by weakly bonding a silver ion to a citrate ion. This novel irrigant was reported to be nontoxic and biocompatible [6,7].

Ozone has numerous beneficial effects, including its antimicrobial activity, the oxidation of bacterial biomolecules and microbial toxins, the ability to remove the smear layer, and opening dentinal tubules to allow the deeper penetration of Ca and fluorine ions into them [8]. Ozonated oils are obtained by the means of chemical reactions that pass pure oxygen and ozone through the oils, and they are potent antibacterial irrigants [9]. Although previous studies have stated the antibacterial efficacy of ozonated olive oil and silver citrate in root canals, this is a pioneer study on its effect on the chemical structure of root dentin.

A variety of chemical irrigating solutions have been studied to determine their modes of action and effectiveness on both the organic and inorganic components of root canal dentin. These effects are directly related to the mechanical, chemical, and physical properties of the dentin structure [10]. Although very limited research has been conducted corelating the effects of various irrigants on root dentin, when considering the long-term success of a root canal treatment, maintaining the chemical and mechanical integrity of root dentin plays an integral role. Thus, it is important that novel irrigants be studied regarding their effects on the dentin microstructure so as to reduce the detrimental and erosive effects of conventional irrigants. This in vitro research aimed to compare and evaluate the effects of three different irrigating solutions on the chemical structure of root canal dentin in extracted human teeth. The null hypothesis states that these irrigating solutions do not alter the chemical structure of root dentin.

2. Materials and Methods

Ethical clearance for the study was obtained before the start of the study from the institutional ethics review committee with code (Ref. BMC&H/I EC/2021-22/32). This in-vitro study was conducted on 48 extracted single-rooted mandibular premolar teeth. They were extracted for the purpose of orthodontic treatment and were collected from the department of oral and maxillofacial surgery.

2.1. Sample Preparation

All collected samples were washed with distilled water and cleaned using ultrasonic scaling. Then, the specimens were stored in a chloramine-T solution. Soft tissues and debris were removed, and a diamond disc (Ray Foster, Huntington Beach, CA, USA) was used to decoronate the tooth at the level of the cemento–enamel junction.

Samples were then randomly assigned to four groups (Gp) with 12 samples each:

Group 1: 5.25% NaOCl;

Group 2: Ozonated olive oil (Ozonoid; Adc Inc. Dentozoneindia; Maharashtra, Mumbai, India);

Group 3: Silver citrate (BioAKT, New Tech Solutions s.r.l., Brescia, Italy);

Group 4: Distilled water.

A conventional access cavity preparation was created to access the root canal system. A size 10 K file was placed in each canal to determine its patency. According to the manufacturer's instructions, the working length was set at 1 mm below the apex, and the canals were progressively enlarged up to size F2 Protaper Gold (Dentsply, Maillefer, Switzerland) while being irrigated with 5 mL of the appropriate irrigating solution in between each instrument, for a total of 20 mL of irrigating solution per tooth sample. The root portion was further sectioned into three parts using a diamond disc, and the middle third of the tooth was used for the experiment, as it presented with adequate root dentin thickness and uniform canal anatomy compared to the apical third. The middle third of the root portion was horizontally cut into slices of 1.5 mm thickness using a diamond disc.

2.2. Fourier Transform Infrared Spectroscopy (FTIR) Analysis

Dentin discs were obtained from the middle third of the root and dentin slices of 1.5 mm thickness was obtained using a diamond disc. Samples were analyzed with an ATR-FTIR (iS50 Nicolet FTIR Spectrometer; ThermoFisher Scientific; Waltham, MA, USA) between 400 and 4000 cm at a 1 cm resolution over the course of ten scans following irrigation with the respective irrigants. Collagen, phosphate, and carbonate peak levels were determined (in cm^{-1}) (Figure 1).

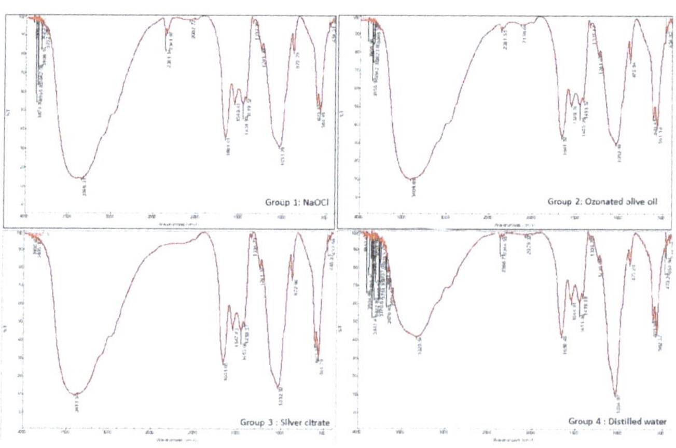

Figure 1. Collagen, phosphate, and carbonate peak levels, as determined by an FTIR spectrometer between 400 and 4000 cm at a 1 cm resolution using ten scans for groups 1, 2, 3, and 4.

2.3. Scanning Electron Microscopy (SEM) and Electron-Dispersive Spectroscopy (EDS) Analysis

The same samples were dried at 37 °C for 48 h, and sample segments were fixed in stubs with the dentin walls upwards. The samples were coated with two thin layers of evaporated carbon in high vacuum by a desk carbon coater and viewed at a magnification of 1500× with a field-emission SEM (JEOL IT-300, JEOL. Ltd., Tokyo, Japan). To measure the atomic percentages of the carbon, oxygen, phosphorus, and calcium levels in the dentin samples, an EDS analysis was carried out in conjunction with SEM using iridium software at an accelerating voltage of 20 KeV (Figure 2).

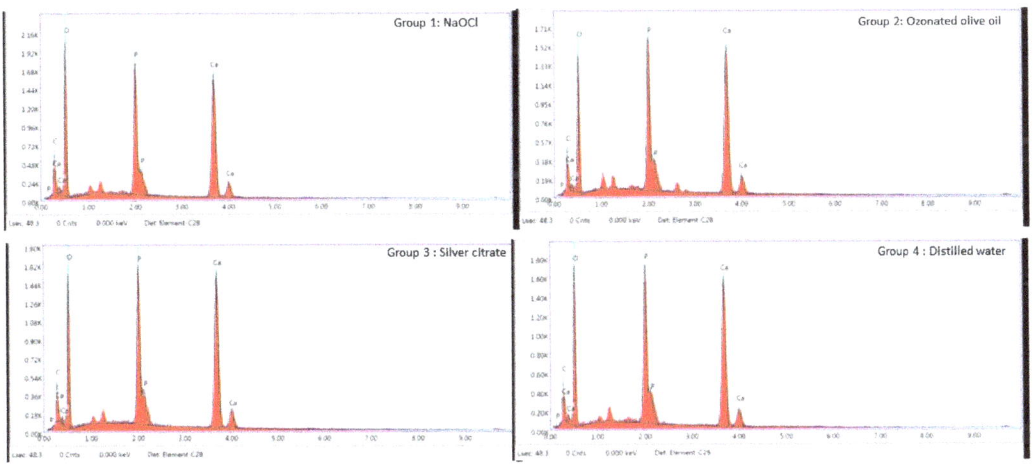

Figure 2. EDS analysis measured atomic percentages of carbon, oxygen, phosphorous, and calcium levels of dentin samples obtained from group 1, 2, 3, and 4 using iridium software at 20 KeV.

2.4. Statistical Analysis

SPSS (Statistical Package For Social Sciences) version 20 (IBM SPASS statistics (IBM Corp. (Armonk, NY, USA)) released 2011) was used to perform the statistical analysis. Inferential statistics such as ANOVA were applied to check the statistical differences in chemical structural changes among the groups, with post hoc Bonferroni for intergroup comparisons.

3. Results

The comparison of the FTIR values among the groups using ANOVA revealed statistically significant differences in the collagen carbonate and phosphate peak values among the groups (Figure 1). A decrease in the collagen level was observed in the NaOCl group compared to the experimental and negative control groups, although the results were not statistically significant ($p > 0.05$). The intergroup comparisons between NaOCl and silver citrate as well as ozonated olive oil revealed significant reductions in the carbonate and phosphate peak values in the NaOCl group ($p < 0.05$) (Figure 3).

The EDS values tabulated for the carbon, oxygen, phosphorous, and calcium peak levels showed significant differences between the groups using an ANOVA (Figure 4). The mean value for the carbon peak levels in NaOCl was significantly different compared to the silver citrate and ozonated olive oil. Silver citrate showed reduced mean values for Ca and P compared to the other groups, although they were not statistically significant ($p > 0.05$) (Figure 4).

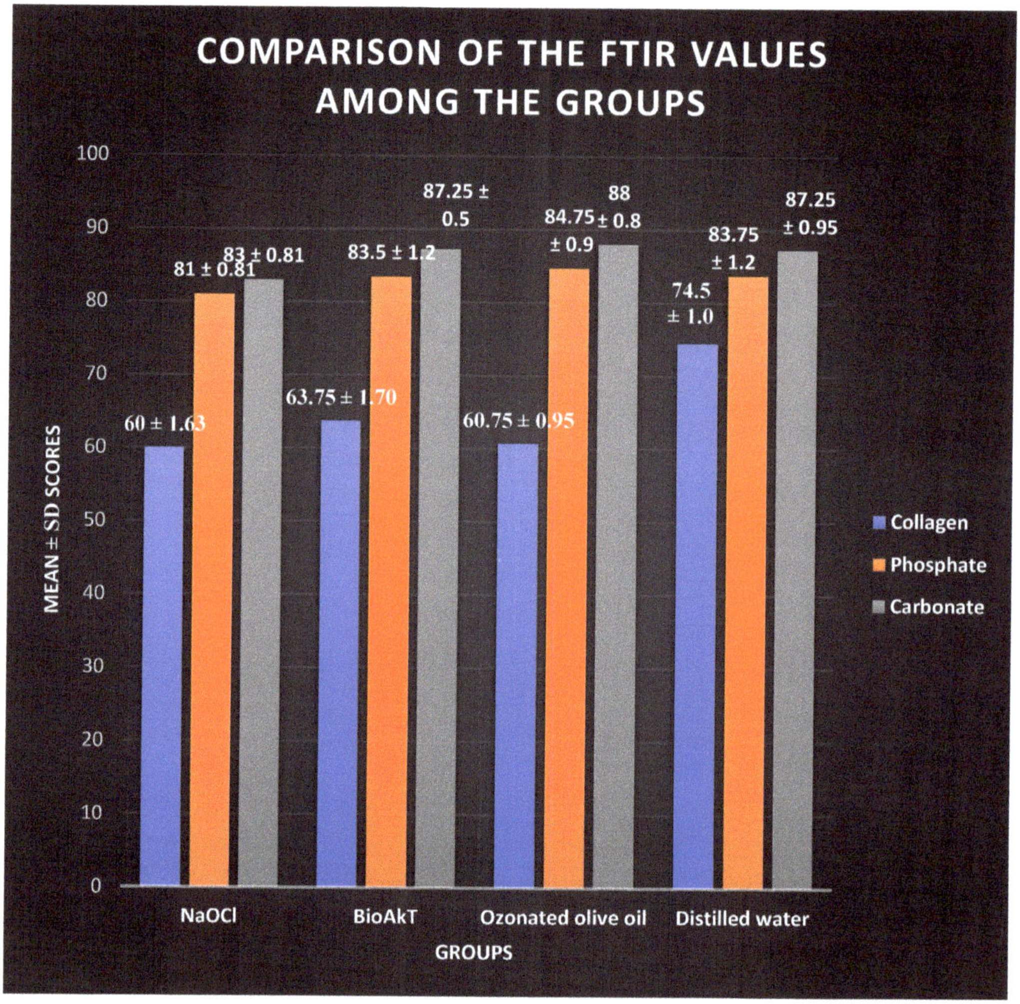

Figure 3. The collagen, phosphate, and carbonate peak values obtained by FTIR analysis for group 1: NaOCl (sodium hypochlorite), group 2: silver citrate, group 3: ozonated olive oil, and group 4: distilled water.

The SEM analysis was performed under 1500× magnification and revealed smear layer removal in the silver citrate group. The partial removal of the smear layer was observed in the sodium hypochlorite group, while ozonated olive oil and distilled water did not exhibit smear layer removal (Figure 5).

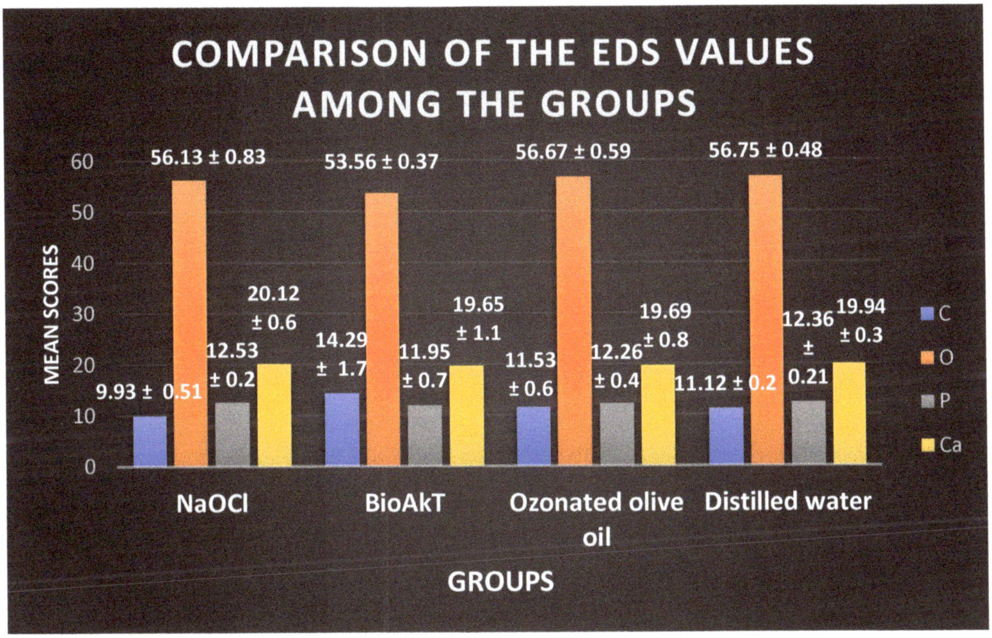

Figure 4. The inorganic mineral content of dentin, as obtained from EDS analysis, after irrigation with the respective irrigants. C—carbon, O—oxygen, P—phosphorous, C—calcium.

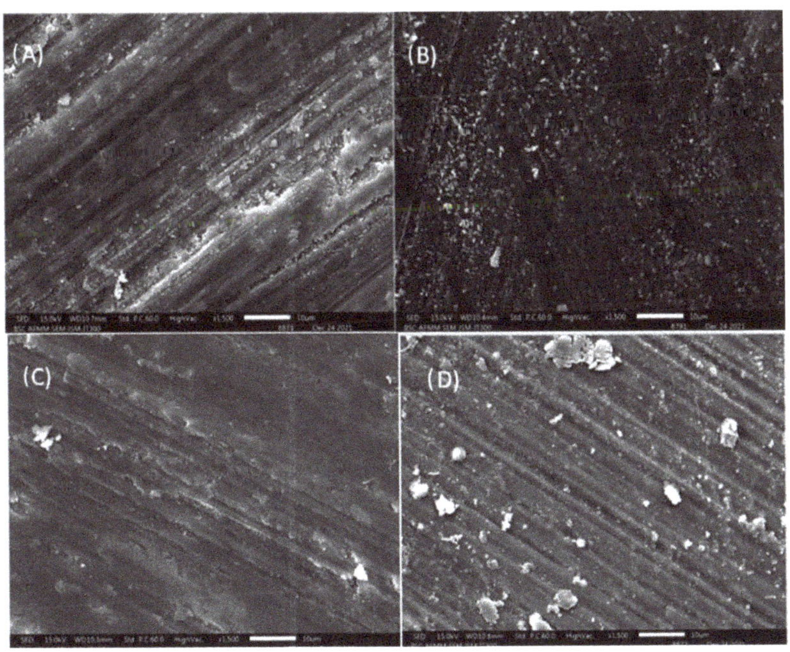

Figure 5. SEM image showing dentin cross sections under 1500× magnification after irrigation with (**A**) sodium hypochlorite, (**B**) ozonated olive oil, (**C**) silver citrate, and (**D**) distilled water.

4. Discussion

Our ability to strike a balance between the "biological" goals and the "mechanical" objectives of the therapy continues to be the clinical challenge of root canal therapy. The ideal irrigation protocol is one that eliminates biofilm for maximum antibacterial effectiveness while having no negative effects on the mechanical integrity of the tooth. The damaged root canal filling–dentin interface during treatment further complicates this requirement. It is important to take into account how endodontic irrigants directly alter the chemomechanical characteristics of the root canal dentin, which has an impact on the effectiveness and durability of all materials used for root canal obturations and restorations.

In the current study, various novel endodontic irrigating solutions were tested individually to determine their effects on dentin through various methods of chemical analysis, including elemental (EDS) and chemical (FT-IR) determinations of the dentin constitution and a surface microstructural analysis by SEM. The null hypothesis was rejected, as various endodontic irrigants used in this study altered the chemical structure of root dentin.

NaOCl significantly reduced the organic components of the dentin such as carbonate and phosphate compared to novel irrigating solutions such as silver citrate and ozonated olive oil. This is in accordance with studies conducted by Sakae et al., who stated that NaOCl is capable of removing magnesium and carbonate ions from dentin [11]. All the irrigating solutions were shown to affect the collagen content of the root dentin, although NaOCl showed the highest mean difference compared to distilled water. NaOCl at a concentration of 1.5% was shown to reduce the collagen content and caused a subsequent reduction in flexural strength, which is in agreement with the observations of this study [12]. At higher concentrations ranging from 5 to 9%, as used in this study, it can cause alterations in the carbon and nitrogen contents of dentin, reducing the dentin microhardness, as reported by Marending et al. The organic-tissue-dissolving properties of sodium hypochlorite as an irrigant on the collagen component of dentin has already been established by various studies, and it has been shown to affect the microhardness of dentin [13–15].

NaOCl spreads on the intrafibrillar water volume of apatite-encapsulated collagen matrix owing to its low molecular weight. Collagen from the "superficial subsurface" of mineralized dentin undergoes oxidative chemical destruction when it comes into contact with sodium hypochlorite. According to Huang et al., dentin specimens can lose their toughness and flexural strength with just a 1 μm depth of collagen degradation on the dentin surface [16].

The silver citrate solution used in the study comprised 4.8% citric acid. Previous studies have stated that citric acid at concentrations between 25% and 50% is an effective endodontic irrigant. However, recent studies have demonstrated the efficacy of citric acid solutions with lower concentrations (<10%) to be operational. In comparison to the other groups, the silver citrate group displayed a slightly larger quantity of calcium dissolution, which is consistent with citric acid's capacity to decalcify hard dental tissues by the chelation of Ca^{2+} ions in a mildly acidic environment [17].

Aldehydes, ketones, and hydrogen peroxide can be generated as a result of the hydrolysis of ozonized oil. As an oxidant, hydrogen peroxide degrades vital biological components such as lipids, proteins, and nucleic acids [18]. Hydrogen peroxide also has an effect on the inorganic components of dentin by acidic demineralization [19]. Silver citrate (BioAKt) exhibited effective smear layer removal compared to the other groups in the study. Several studies have stated the efficiency of citric acid in smear layer removal, which was superior to EDTA at similar concentrations [20]. It has been suggested that irrigant solutions with the ability to remove the smear layer may reduce dentin microhardness; however, when these solutions are used for a brief period of time inside the root canal, this reduction does not appear to have a negative impact on the fracture resistance of teeth that have undergone endodontic treatment [21]. All irrigating solutions used in this study were only briefly in contact with the dentin, which may account for the comparable mineral compositions of the dentin seen in the experimental groups (Figure 4). Furthermore, the exposure time of irrigants to the dentinal wall has been standardized, as it could have an impact on the elemental composition of carbon, calcium, phosphorous, and oxygen [10].

The mechanical, physical, and chemical features of root dentin must be conserved against any harmful impacts of chemical substances [22]. However, it should be emphasized that this study has some limitations that should be addressed. One of the main limitations of this study is that it was conducted in vitro, thus failing to simulate the conditions of the oral cavity. Future research is needed to evaluate the changes in physical properties upon the usage of these irrigants and to correlate them with the results of this study. In addition, further research with the criteria used in this study are encouraged to emphasize the relevance of the findings.

5. Conclusions

It was apparent that all studied irrigation sequences potentially result in some alteration in the inorganic and organic contents of the dentin. Furthermore, it can also be concluded from the results that the silver citrate solution and the ozonated olive oil cause fewer changes in the organic and mineral contents of dentin than sodium hypochlorite.

Author Contributions: Conceptualization, I.P., A.M. and M.I.K.; methodology, A.M. and M.I.K.; software, A.K.S.; validation, D.H., R.K.N. and A.K.S.; formal analysis, A.R.; investigation, D.H., R.K.N. and A.K.S.; resources, M.B.M. and S.N.B.; data curation, D.H., R.K.N. and A.K.S.; writing—original draft preparation, I.P., D.H. and M.I.K.; writing—review and editing, A.R., M.I.K. and G.A.S.; visualization, M.B.M. and S.N.B.; supervision, M.I.K. and G.A.S.; project administration, A.M.; funding acquisition, G.A.S. All authors have read and agreed to the published version of the manuscript.

Funding: This research received no external funding.

Institutional Review Board Statement: The investigations were carried out following the rules of the Declaration of Helsinki of 1975. Ethical clearance for the study was obtained before the start of the study from the ethics review committee of SJM Dental College And Hospital, Chitradurga, Karnataka, India, with code Ref. BMC&H/I EC/2021-22/32.

Informed Consent Statement: Not applicable.

Data Availability Statement: The data will be made available by the corresponding author on reasonable request.

Conflicts of Interest: The authors declare no conflict of interest.

References

1. Saleh, A.A.; Ettman, W.M. Effect of endodontic irrigation solutions on microhardness of root canal dentine. *J. Dent.* **1999**, *27*, 43–46. [CrossRef]
2. Kaur, P. Role of Irrigants in Endodontics. *J. Dent. Probl. Solut.* **2020**, *7*, 100–104.
3. Ghorbanzadeh, S.; Loodaricheh, S.A.; Samizade, S.; Zadsirjan, S. Irrigants in endodontic treatment. *Structure* **2015**, *11*, 12.
4. Cárdenas-Bahena, Á.; Sánchez-García, S.; Tinajero-Morales, C.; González-Rodríguez, V.M.; Baires-Várguez, L. Use of sodium hypochlorite in root canal irrigation. Opinion survey and concentration in commercial products. *Rev. Odontol. Mex.* **2012**, *16*, 252–258.
5. Haapasalo, M.; Shen, Y.; Qian, W.; Gao, Y. Irrigation in endodontics. *Dent. Clin. N. Am.* **2010**, *54*, 291–312. [CrossRef]
6. Generali, L.; Bertoldi, C.; Bidossi, A.; Cassinelli, C.; Morra, M.; Del Fabbro, M.; Savadori, P.; Ballal, N.V.; Giardino, L. Evaluation of cytotoxicity and antibacterial activity of a new class of silver citrate-based compounds as endodontic irrigants. *Materials* **2020**, *13*, 5019. [CrossRef] [PubMed]
7. Tonini, R.; Giovarruscio, M.; Gorni, F.; Ionescu, A.; Brambilla, E.; Mikhailovna, I.M.; Luzi, A.; Maciel Pires, P.; Sauro, S. In vitro evaluation of antibacterial properties and smear layer removal/sealer penetration of a novel silver-citrate root canal irrigant. *Materials* **2020**, *13*, 194. [CrossRef] [PubMed]
8. Shoukheba, M.Y.; Ali, S.Y.; Gamal, S.M. Ozonated olive oil gel versus nano-silver mouthwash in root biomodification: A scanning electron microscopic study. *J. Int. Oral Health* **2021**, *13*, 298.
9. Wassef, N.M.; Fouad, M.A. Antibacterial effect of ozonated olive oil versus sodium hypochlorite against Enterococcus faecalis bacteria: In vitro study. *Egypt. Dent. J.* **2019**, *65*, 3271–3276. [CrossRef]
10. Rath, P.P.; Yiu, C.K.; Matinlinna, J.P.; Kishen, A.; Neelakantan, P. The effect of root canal irrigants on dentin: A focused review. *Restor. Dent. Endod.* **2020**, *45*, e39. [CrossRef]

11. Sakae, T.; Mishima, H.; Kozawa, Y. Changes in bovine dentin mineral with sodium hypochlorite treatment. *J. Dent. Res.* **1988**, *67*, 1229–1234. [CrossRef] [PubMed]
12. Bosaid, F.; Aksel, H.; Makowka, S.; Azim, A.A. Surface and structural changes in root dentine by various chelating solutions used in regenerative endodontics. *Int. Endod. J.* **2020**, *53*, 1438–1445. [CrossRef]
13. Marending, M.; Luder, H.U.; Brunner, T.J.; Knecht, S.; Stark, W.J.; Zehnder, M. Effect of sodium hypochlorite on human root dentine–mechanical, chemical and structural evaluation. *Int. Endod. J.* **2007**, *40*, 786–793. [CrossRef] [PubMed]
14. Cohen, S.; Stewart, G.G.; Laser, L.L. The effects of acids, alkalines and chelating agents on dentine permeability. *Oral Surg. Oral Med. Oral Pathol. Oral Radiol. Endod.* **1970**, *29*, 631–634. [CrossRef]
15. Gordon, T.M.; Damato, D.; Christner, P. Solvent effect of various dilutions of sodium hypochlorite on vital and necrotic tissues. *J. Endod.* **1981**, *7*, 466–469. [CrossRef]
16. Gu, L.S.; Huang, X.Q.; Griffin, B.; Bergeron, B.R.; Pashley, D.H.; Niu, L.N.; Tay, F.R. Primum non nocere–The effects of sodium hypochlorite on dentin as used in endodontics. *Acta Biomater.* **2017**, *61*, 144–156. [CrossRef] [PubMed]
17. Mohammadi, Z.; Shalavi, S.; Yaripour, S.; Kinoshita, J.I.; Manabe, A.; Kobayashi, M.; Giardino, L.; Palazzi, F.; Sharifi, F.; Jafarzadeh, H. Smear Layer Removing Ability of Root Canal Irrigation Solutions: A Review. *J. Contemp. Dent. Pract.* **2019**, *20*, 395–402. [CrossRef] [PubMed]
18. David, A.W.; Satish, S.W.; Dole, S.; Kovvuru, S.K.; Gowda, B. A Comparative Evaluation of the Effectiveness of Saline, Chlorhexidine and Ozone Irrigating Solutions on Microorganisms in the Root Canal—An In Vivo Study. *Acta Sci. Dent. Sci.* **2020**, *4*, 76–81. [CrossRef]
19. Pécora, J.D.; Cruzfilho, A.M.; Sousaneto, M.D.; Silva, R.G. *In-vitro* action of various bleaching agents on microhardnesss of dentin. *Braz. Dent. J.* **1994**, *5*, 129–334. [PubMed]
20. Zehnder, M.; Schmidlin, P.; Sener, B.; Waltimo, T. Chelation in root canal therapy reconsidered. *J. Endod.* **2005**, *31*, 817–820. [CrossRef] [PubMed]
21. Oelichmann, J. Surface and depth-profile analysis using FTIR spectroscopy. *Fresenius' Z. Anal. Chem.* **1989**, *333*, 353–359. [CrossRef]
22. Yassen, G.H.; Eckert, G.J.; Platt, J.A. Effect of intracanal medicaments used in endodontic regeneration procedures on microhardness and chemical structure of dentin. *Restor. Dent. Endod.* **2015**, *40*, 104–112. [CrossRef] [PubMed]

Journal of
Functional Biomaterials

MDPI

Article

Cytotoxicity of V-Prep Versus Phosphoric Acid Etchant on Oral Gingival Fibroblasts

Victor Ghoubril [1,*], Sylvie Changotade [2], Didier Lutomski [2], Joseph Ghoubril [1], Carole Chakar [3], Maher Abboud [4], Louis Hardan [5], Naji Kharouf [6,*] and Elie Khoury [1,*]

1. Department of Orthodontics, School of Dentistry, Saint-Joseph University, Beirut 1107 2180, Lebanon
2. Unité de Recherches Biomatériaux Innovants et Interfaces, URIT, Université Sorbonne Paris Nord—Université de Paris, 93017 Paris, France
3. Department of Periodontology, School of Dentistry, Saint-Joseph University, Beirut 1107 2180, Lebanon
4. Unité Environnement Génomique et Protéomique, U-EGP, Faculté des Sciences, Université Saint-Joseph, Campus des Sciences et Technologies Mar Roukos-B.P. 1514, Riad El Solh, Beirut 1107 2050, Lebanon
5. Department of Restorative Dentistry, School of Dentistry, Saint-Joseph University, Beirut 1107 2180, Lebanon
6. Department of Biomaterials and Bioengineering, INSERM UMR_S 1121, Biomaterials and Bioengineering, 67000 Strasbourg, France

* Correspondence: victor.ghoubril1@usj.edu.lb (V.G.); dentistenajikharouf@gmail.com (N.K.); elie.khoury@usj.edu.lb (E.K.); Tel.: +33-667522841 (N.K.)

Abstract: The most used etchant in dental daily practice is the phosphoric acid (P.A.; 37%). However, acid etchants can induce necrosis on the oral mucosa and cause the ulceration of periodontal tissue when a rubber dam is not used. V-prep is a new practical alternative, and it has satisfactory results. It is used as a preparation before the application of a resin-modified glass ionomer composite (RMGIC) to bond the orthodontic brackets. The aim of this study was to investigate the effect of the V-prep on oral gingival fibroblasts cells by comparing the cell damage and cell viability after the use of V-prep and a conventional phosphoric acid etchant with different application times and concentrations. Therefore, Gingival fibroblasts passage 6 (GFP6) was grown and treated with an acid etchant and V-prep at three different concentrations (1:1, 1:2 and 1:10) for two different application durations (30 s and 1 min). The morphological changes, cell death and cell viability were assessed. Pyknosis, karyolysis, nucleus reversible and irreversible damages and membrane destruction were observed for both of the etchants at the higher concentrations and longer application durations. Mann–Whitney U-tests were used for the statistical analyses. The application of the V-prep for 30 s showed better values than the acid etchant did in the cell damage analysis and cell viability analysis ($p = 0.03$). V-prep at a 1:10 concentration applied for a 30 s duration can preserve the viability of gingival fibroblasts cells up to 100%. The toxicity of V-prep is equal or lower than the toxicity of the acid etchant that is commonly used in dentistry. Thus, the V-prep can be used with precautions intra-orally, and it should be applied on the enamel as a gel for 30 s only before it is rinsed and removed.

Keywords: cell viability; dental acid etchants; oral mucosa; phosphoric acid; V-prep

1. Introduction

Bracket bonding in orthodontic treatments requires a fast effective procedure with a high bond strength to resist the orthodontic forces and masticatory loads so that no bracket–enamel failure occurs [1–3]. Therefore, the conventional technique [4] consists of a preparation with the phosphoric acid (PA) etchant (37%) for 30 s, water rinsing, drying and bonding the brackets with a composite resin [5]. Universal adhesives with alternative techniques have been demonstrated to enhance the bond strength on the enamel and on the dentin in oral cavity restorations [6]. The drawbacks on this bonding procedure are the white spots that are encountered at the end of the treatment and the sticking on the surfaces with gingival fluid or blood excretion [7]. Another approach in the bonding procedure using a resin-modified glass ionomer cement (RMGIC) has been tested, and it has showed good results on wet surfaces, enhancing the chemical bond and releasing the fluoride during the treatment to help protect the tooth [8]. The only disadvantage was the shear bond strength of the material when it was compared to the conventional technique [9].

A previous study, which was published by the same author, has introduced a preparation procedure with V-prep before using the RMGIC. The results showed a non-significant difference in the shear bond strength (SBS) when it was compared to the PA and the composite resin [10]. To minimize the tooth damage without compromising the adhesive performance, most of the manufacturers of dental acid etchants have recommended from 15 to 30 s of use when the PA is between 32% and 40% [11]. Inadequate rinsing or remaining dental acid etchants can cause chemical burning, irritation, intra and extra-oral inflammation [12,13]. PA (37%) can lead to necrosis in the oral mucosa and ulcerative lesions of the periodontal tissue [14]. The bracket-bonding procedure for orthodontic treatments is performed without a rubber dam, while studies have demonstrated that it can protect the gingival tissue while bonding ceramic veneers [15]. V-prep has been mentioned to contain PA and sodium hypochlorite in dilution with a gel for easier manipulation.

V-prep manipulation is similar to the conventional PA etchants as mentioned in another study [10], and should be applied for 15 to 30 s. Then, the surface should be adequately rinsed before using the RMGIC as a bonding product instead of the composite resin. Both of the etchants are applied on the bonding area of the enamel without a rubber dam protection [16]. The removal process is performed with a suction tip and high water flow, rinsing in order to dilute the etchant gel (according to the American Association of Orthodontists recommendations). Gingival fibroblasts are the first surrounding tissue to become in contact with the diluted etchant in different concentrations. Studies comparing the effect of the acid etchant on gingival fibroblasts tested three concentrations of contact, 1:1 (undiluted), 1:2 and 1:10 [17]. The toxicology and the damage caused on the oral cells have not been tested for V-prep. Cytotoxicity is essential before the use of the product in vivo [18]. Therefore, the aim of this study was to investigate the effect of V-prep on the gingival fibroblasts cells by comparing the cell damage and cell viability after the use of V-prep and a conventional phosphoric acid etchant with different application times and concentrations.

2. Materials and Methods

This study was performed at URIT of Université Sorbonne Paris Nord. The study has been approved by the ethical committee of Saint-Joseph University of Beirut (USJ-2020-010).

2.1. Cell Cultures

Normal human gingival fibroblasts (GF), which were obtained by surgical periodontal operation, were grown in Dulbecco's Modified Eagles Medium (Gibco BRL, Grand Island, NY, USA) supplemented with 10% fetal bovine serum and 1% penicillin/streptomycin for 6 passages. The cells were maintained in an incubator at 37 °C with an atmosphere of 5% CO_2. The cell culture medium was changed every 3 days. The cells were detached by an enzymatic treatment with trypsin EDTA (Invitro gen, life technology) and seeded in 24-well plates at a rate of 1×10^5. After 24 h, the cells were left in a culture that was untreated or treated with different dilutions (undiluted, 1:2 or 1:10) of orthophosphoric acid (Medental, Vista, CA, USA) or V-prep (concept product described in a previous article [10]) for 30 s or 1 min.

2.2. Giemsa Staining

To observe the morphological changes caused by orthophosphoric acid etchant and the V-prep, the cells were rinsed with PBS (Gibco, life technology, Grand Island, NY, USA), then, they were fixed according to the following procedure: absolute ethanol/PBS (50/50) at 4 °C for 5 min. Then, after 5 min of rehydration, the cells were then stained with Giemsa (Labonord, MercK, Rahway, NJ, USA) according to the supplier's recommendations. The cells were observed at 100–250-fold magnification using an optical microscope (ZEISS AXIOPLAN, Jena, D-07740, Germany). The nucleus appeared to be deep purple, and the cytoplasm appeared be to brown or pink by light microscopy. Cell damage contains both the irreversible cell injuries, including karyorrhexis, pyknosis, karyolysis and membrane destruction, and the reversible cell injuries, including vacuole and cell swelling (enlargement).

2.3. Cell Death Quantification

To quantify death cell after exposition to orthophosphoric acid etchant and V-prep, the treated cells (1×10^5) with the same procedure of concentration and time for orthophosphoric acid etchant and V-prep were counted using the Beckman Coulter Vi-CELL XR (Brea, CA, USA) after Trypsin EDTA treatment to ensure the total remaining cells' transportation into the device. The percentage of dead cells was calculated by counting the total number of cells.

2.4. Cytotoxicity Assay

To identify the cytotoxic effect of dental acid etchants and V-prep on gingival fibroblast cell, a 3-(4,5-dime-thylthiazol-2-yl)-2,5-diphenyltetrazolium bromide (MTT) assay was performed. In brief, the cells (1×10^5) were seeded in to 24-well plates and different concentration of orthophosphoric acid etchant (non-treated, undiluted, dilution ratios of 1:2 and 1:10, respectively) and V-prep (non-treated, undiluted, dilution ratios of 1:2 and 1:10, respectively) were applied for 30 s and 1 min (Table 1). After the cell stabilization for 24 h, MTT at 0.05 mg/mL (Sigma Aldrich, Burlington, MA, USA) was added to each well and incubated for 4 h at 37 °C. After removing the MTT solution, dimethyl sulfoxide (Merck, Rahway, NJ, USA) was added to dissolve the formazan dye crystals. The optical density was measured at a wavelength of 570 nm and a reference wavelength of 680 nm using a microplate reader (Asys UVM340, Biochrom, Cambridge, UK). The percentage of cell viability was calculated using the following formula:

$$\%\text{Viability} = \frac{Mean\ OD\ sample}{Mean\ OD\ untreated\ cells} \times 100 \quad (OD:\ Optical\ density) \qquad (1)$$

Table 1. 24-well plate treated for 30 s using V-prep and acid etchant in different concentrations before MTT procedure.

30 s	1	2	3	4	5	6
A		V-prep 1:1 (undiluted)			Acid etch 1:1 (undiluted)	
B		V-prep 1:2			Acid etch 1:2	
C		V-prep 1:10			Acid etch 1:10	
D		Non-treated			Non-treated	

2.5. Statistical Analysis

All of the statistical analyses were performed by SPSS ver. 26.0 (IBM Corp., Armonk, NY, USA). Mann–Whitney U-tests were used to compare between the control and experimental groups. Each test was performed at least in triplicate. The results were reported as the mean ± standard deviation. A value of $p < 0.05$ was considered to be statistically significant.

3. Results

3.1. Morphological Changes Gingival Fibroblasts

The cells were treated with V-prep and acid etchant for 30 s and 1 min at three different concentrations, 1:1, 1:2 and 1:10. Then, Giemsa staining was performed. The same procedure was also performed on a control group in which the cells were not treated. The examinations and observations were performed using a ZEISS microscope at 100–250-fold magnification to identify the cellular changes (Figures 1 and 2).

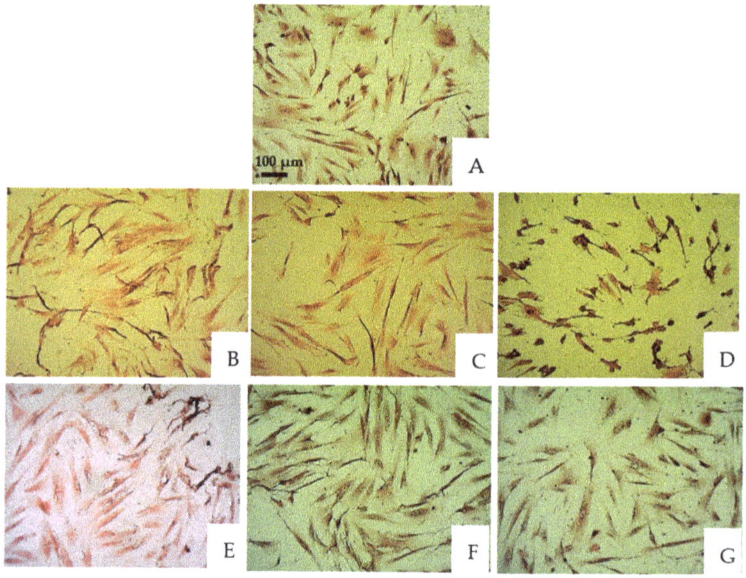

Figure 1. Microscope observations of oral epithelial cells stained by Giemsa under 100-fold magnification. (**A**) Control group. (**B**) V-prep, 30 s, undiluted. (**C**) V-prep, 30 s, 1:2. (**D**) V-prep, 30 s, 1:10. (**E**) Acid etchant, 30 s, undiluted. (**F**) Acid etchant, 30 s, 1:2. (**G**) Acid etchant, 30 s, 1:10.

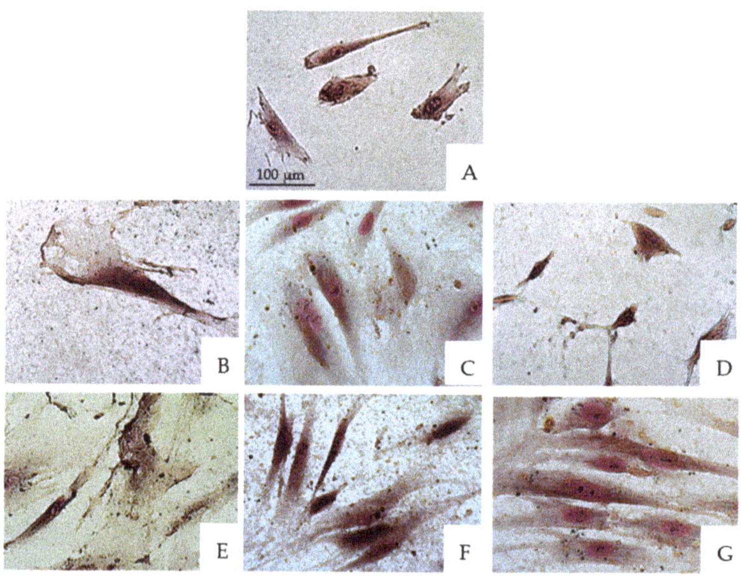

Figure 2. Microscope observations of oral epithelial cells stained by Giemsa under 250-fold magnification. (**A**) Control group. (**B**) V-prep, 30 s, undiluted. (**C**) V-prep, 30 s, 1:2. (**D**) V-prep, 30 s, 1:10. (**E**) Acid etchant, 30 s, undiluted. (**F**) Acid etchant, 30 s, 1:2. (**G**) Acid etchant, 30 s, 1:10.

Similar pictures were taken for the microscope observations in the 30 s and 1 min application tests of the etchants, with there being no big differences. Figures 1 and 2 show, respectively, the microscope observations of the oral epithelial cells stained by Giemsa under 100-fold and 250-fold magnifications with a 30 s application of the etchant. The acid etchant and V-prep caused remarkable cell damage. V-prep at a 1:10 ratio showed normal cells division when it was used for 30 s (D) when it was compared to the control group with the untreated cells (A). Pyknosis and karyolysis were observed in the low concentrations of the acid etchant (G). Cell injury in the nucleus started to be observed at a concentration of 1:2 (C,F). When the undiluted etchants were used, the percentage of cell damage was increased with an enlarged nuclear injury and subsequent membrane destruction (B,E). Cell damage contains both the irreversible cell injuries, including karyorrhexis, pyknosis, karyolysis and membrane destruction, and the reversible cell injuries, including vacuole and cell swelling.

3.2. Cell Death Quantification

The percentage of dead cells was calculated by counting the total number of cells for each condition.

Gingival fibroblasts treated with the acid etchant for 30 s showed a significant higher death ratio ($p < 0.001$) when they were compared to the non-treated cells. The cells treated with V-prep at 1:1 (undiluted) and 1:2 for 30 s showed a significant death ratio ($p < 0.001$). Only the V-prep at a 1:10 concentration did not kill the cells significantly ($p = 0.097$) when it was used for 30 s (Table 2). The cell death rate dropped by 2.82 folds (33.58%) and by 2.35 folds (47.8%) for the V-prep and acid etchant conditions, respectively, between a 1:1 concentration and a 1:10 concentration when they were used for 30 s (Figure 3).

Table 2. p-value using Mann–Whitney U test for cell death analyses after 30 s of different etchant application.

p-Value	Non-Treated	Vprep 1:1	Vprep 1:2	Vprep 1:10	Ac 1:1	Ac 1:2	Ac 1:10
Non-treated		0.000	0.000	0.097	0.000	0.000	0.000
Vprep 1:1	0.000		0.922	0.035	0.048	0.820	0.060
Vprep 1:2	0.000	0.922		0.037	0.045	0.810	0.061
Vprep 1:10	0.097	0.035	0.037		0.002	0.032	0.047
Ac 1:1	0.000	0.048	0.045	0.002		0.055	0.011
Ac 1:2	0.000	0.820	0.810	0.032	0.055		0.058
Ac 1:10	0.000	0.060	0.061	0.047	0.011	0.058	

The highest value (83.25%) for the dead gingival fibroblast cells was observed for the application of undiluted acid etchant (30 s). The one minute application has showed lower values of cell death that did not exceed 37%, while the non-treated cells showed a 13% death, and a significant difference has been observed between the treated and non-treated cells for 1 min of the etchant application ($p < 0.05$). However, non-significant differences were observed between the application of acid etchant and V-prep with 1:1, 1:2 and 1:10 concentrations for 1 min ($p = 0.8$; $p = 0.77$; $p = 0.23$) (Table 3). A slight decrease in the number of dead cells of 5% and 4% were noted when the V-prep and acid etchant concentrations, respectively, were lower at 1:1 and 1:10 (Figure 4).

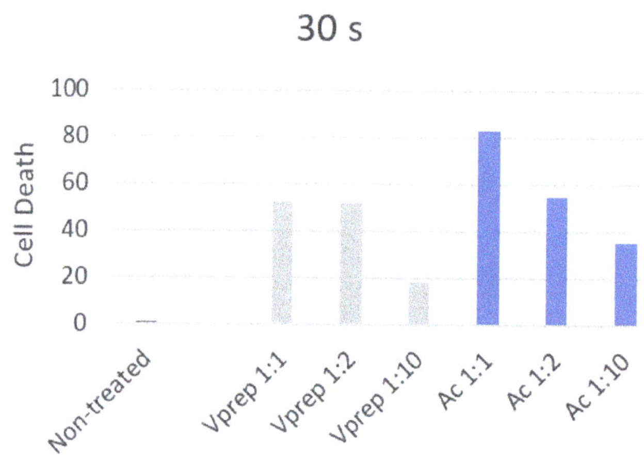

Figure 3. Cell death percentage after 30 s of different etchant application.

Table 3. p-value using Mann–Whitney U test for cell death analyses after 1 min of different etchant application.

p-Value	Non-Treated	Vprep 1:1	Vprep 1:2	Vprep 1:10	Ac 1:1	Ac 1:2	Ac 1:10
Non-treated		0.000	0.012	0.012	0.012	0.014	0.015
Vprep 1:1	0.010		0.800	0.800	0.800	0.350	0.120
Vprep 1:2	0.012	0.800		0.880	0.880	0.770	0.120
Vprep 1:10	0.012	0.800	0.880		0.860	0.550	0.230
Ac 1:1	0.012	0.800	0.880	0.860		0.550	0.230
Ac 1:2	0.014	0.350	0.770	0.550	0.550		0.470
Ac 1:10	0.015	0.120	0.120	0.230	0.230	0.470	

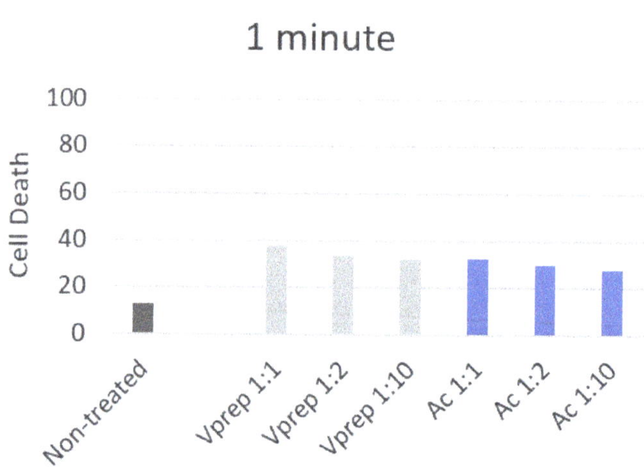

Figure 4. Cell death percentage after 1 min of different etchant application.

3.3. Cell Viability Analysis

MTT assays were performed to investigate the cell viability after the application of the acid etchant and V-prep for 30 s and 1 min at different concentrations 1:1, 1:2 and 1:10. The MTT was then removed, and the cells were regrown in fresh culture for 24 h. To assess the minimum and the maximum cell viability percentages, an application of dimethyl sulfoxide (DMSO) was applied on one of the non-treated control groups (positive control), while the others were left without having any application (negative control).

All of the 30 s and 1 min cell viability values showed a decrease by a minimum of three folds when they were treated except the V-prep 1:10 treatment. The cell viability has not changed between the untreated control group and the V-prep 1:10 group when it was applied for 30 s or 1 min.

At 30 s, the undiluted V-prep showed a higher percentage (32.8%) than the undiluted acid etchant did (22.4%) with a significant difference ($p = 0.03$) (Table 4). The V-prep at a 1:2 ratio (33%) had a cell viability value that was significantly higher than that of the acid etchant at a 1:2 ratio (25%) ($p = 0.04$) (Figure 5).

Table 4. p-value using Mann–Whitney U test for cell viability analyses after 30 s of different etchant application.

p-Value	Non-Treated	Vprep 1:1	Vprep 1:2	Vprep 1:10	Ac 1:1	Ac 1:2	Ac 1:10
Non-treated		0.000	0.000	0.999	0.000	0.000	0.000
Vprep 1:1	0.000		0.880	0.000	0.030	0.035	0.034
Vprep 1:2	0.000	0.880		0.000	0.033	0.040	0.060
Vprep 1:10	0.999	0.000	0.000		0.000	0.000	0.000
Ac 1:1	0.000	0.030	0.033	0.000		0.072	0.053
Ac 1:2	0.000	0.035	0.040	0.000	0.072		0.058
Ac 1:10	0.000	0.034	0.060	0.000	0.053	0.058	

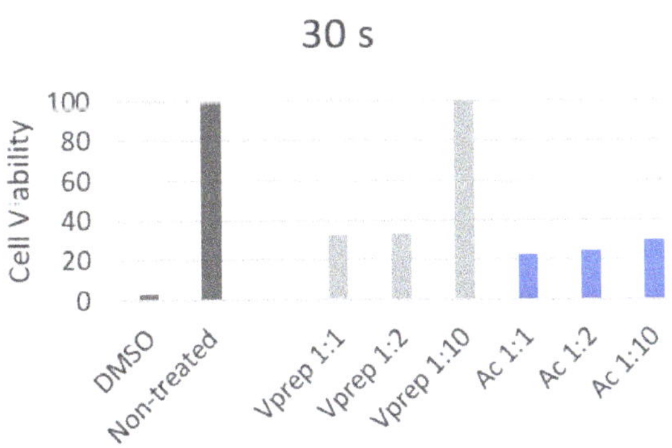

Figure 5. Cell viability percentage after 30 s of different etchant application.

At 1 min, the undiluted V-prep showed a lower percentage (19.5%) than the undiluted acid etchant did (31%) with a significant difference ($p = 0.028$) (Table 5). There was no significant difference between the values of the V-prep at a 1:2 ratio (27%) and the acid etchant at a 1:2 ratio (29%) (Figure 6).

V-prep at a 1:10 ratio preserved the cells viability at a 100% rate with a very significant difference compared to the acid etchant (1:10) value (30%) ($p < 0.001$) for both of the application durations, 30 s and 1 min.

Table 5. p-value using Mann–Whitney U test for cell viability analyses after 1 min of different etchant application.

p-Value	Non-Treated	Vprep 1:1	Vprep 1:2	Vprep 1:10	Ac 1:1	Ac 1:2	Ac 1:10
Non-treated		0.000	0.000	0.999	0.000	0.000	0.000
Vprep 1:1	0.000		0.044	0.000	0.028	0.034	0.033
Vprep 1:2	0.000	0.044		0.000	0.062	0.070	0.068
Vprep 1:10	0.999	0.000	0.000		0.000	0.000	0.000
Ac 1:1	0.000	0.028	0.062	0.000		0.080	0.081
Ac 1:2	0.000	0.034	0.070	0.000	0.080		0.400
Ac 1:10	0.000	0.033	0.068	0.000	0.081	0.400	

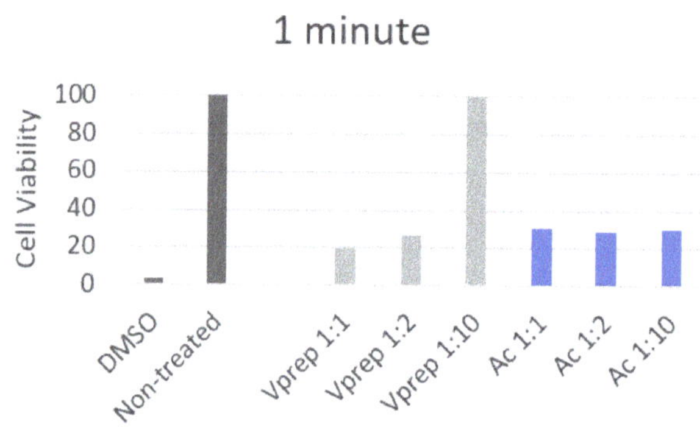

Figure 6. Cell viability percentage after 1 min of different etchant application.

4. Discussion

Clinical research and studies have examined the effects of dental acid etchants on human tissues [17]. Necrosis and chemical burns were reported on the gingiva, the facial skin and the tongue [19]. Acid etchants containing 37% phosphoric acid, is commonly used in restoration treatments and in orthodontics [5]. V-prep is also a new etchant that has been tested in vitro with good shear bond strength (SBS) values when it is used with resin modified glass ionomer composite RMGIC when it is compared the values of the conventional composite bonding material prepared by an acid etchant [10]. Previous studies showed that RMGIC has many advantages over the conventional composite resin [8]. The toxicity of the V-prep preparation before bonding with RMGIC is essential in order to study the combination bonding procedure in vivo [18]. With the results of our study, our hypothesis finds that the toxicity of V-prep is similar or better when it is compared to the acid etchants that are commonly used is accepted.

In this study, the cytotoxicity tests showed similar or better effects on the gingival fibroblasts when we were comparing the V-prep to the acid etchant. Morphological changes, cell death and cell viability were investigated for acid etchant and V-prep at different concentrations (1:1, 1:2 and 1:10) and at two application durations (30 s and 1 min) on the gingival cells. First, the microscope observations showed changes between the treated and non-treated cells, such as pyknosis, karyolysis, membrane destruction, reversible and irreversible cell injuries, vacuole and cell swelling [20]. The damage was observed to be higher with the undiluted etchants without notable differences between the acid etchant and V-prep when they were used at the same concentration and for the same application time.

Secondly, cell death was assessed in values and percentages [21]. The acid etchant showed higher values of cell death when it was compared with V-prep at the 30 s application time. The values were decreased when the etchant was diluted. The lowest value was recorded when the V-prep was applied for 30 s at a 1:10 concentration. When it was used for 1 min, all of the values were lower. The cell death quantification reports the ratio of cells that were penetrated by the colorant (trypan blue) over the total number of cells via the membrane. It does not take into consideration if the damage is reversible or irreversible which leads us to think that there being lower values at 1 min compared to 30 s are due to the gingival cells with reversible damage which occurred in the first 30 s, only.

Thirdly, the gingival cells were analysed by viability because the cell damage is different from the potential of viability [22]. Cell damage can be reversible with slight transformations and modifications, while the cell can still be able to renew and regenerate. This leads us to say that the most accurate test is the cell viability test. The values showed that cell viability was decreased by a minimum of three folds when we used the acid etchant or V-prep. The harm was minimized when the etchant was diluted but not significantly, except for the V-prep at a 1:10 ratio. The application of the V-prep for 30 s or 1 min is harmless, and it keeps the percentage of viability at 100%.

The results of this study regarding the acid etchant application at different concentrations and durations are in complete accordance with similar studies [17,23,24]. Kim et al. found that the acid etchant can damage the vacuoles and the nucleus of the cells when it is applied for 10 s or more at a concentration of 18.5% (referring to 1:2 in our study). Accordingly, they also found that the cell viability was significantly reduced even when the acid etchant was applied for less than 30 s at a concentration of 1:10 [17]. Frob et al. found that both self-etchant adhesive and etchant-rinse adhesive methods are toxic on the normal human gingival fibroblasts [23]. Pupo et al. observed shrinkage and damage in the cells, and they identified the dental etchants to have an increased toxicity on the gingival cells [24]. All of the studies mentioned with similar results recommend in their conclusion, a wise use of the dental etchants, taking into consideration their toxicity and the manufacturer guidelines. Understanding the use of each adhesive method and its advantages over the others is necessary [25].

Practically, when it was used, the acid etchant was applied undiluted at a 1:1 ratio for 30 s before rinsing with a high flow of water for another 30 s. The next step was drying and applying the adequate bonding material in order to fix the orthodontic brackets. During the rinsing with water, the etchant is highly diluted, and the effect is reduced. The dental etchant is only used as a gel texture in order to be able to control the surface of application and limit the damage of the oral tissues [26]. However, the diluted etchant can hit the oral epithelial cells before its complete removal. Thus, better cell viability values at a 1:10 concentration are encouraging and can be a high advantage for the use of V-prep over the acid etchant.

This study encountered few limitations in the statistical analyses. The extracted values on the counting machine when we one is identifying the cell death can vary from one count to another. Therefore, the repetition of each count was required. Moreover, the action of the etchant can be concentrated differently in the same culture of cells, challenging the collection of data results. The findings of this study showed a high toxicity rate of the acid etchant application, which is in accordance with other studies results. Thus, the conventional etching technique becomes questionable. A clinical study on the comparison of the V-prep and the acid etchant can be a step forward in the evolution of bracket bonding in orthodontics.

5. Conclusions

The toxicity of the V-prep showed similar or lower values when it was compared to the acid etchant. However, the acid etchant is used with recommended precautions as a preparation before bracket bonding in orthodontics. With better toxicity values, V-prep can be a substitute to acid etchants.

All of the dental etchants can damage the oral epithelial cells at all concentrations and application time durations when a rubber dam is not used. The use of any etchant should be in a gel texture to control the surface of application and limit the spread on the enamel to save the other epithelial cells from damage.

V-prep requires a gel texture to control the surface of application. It should be applied undiluted for a maximum of 30 s, and it should be well rinsed just after to reach a 10 times dilution before the contact with other oral tissues. The removal should be performed as it is conducted with the acid etchant.

Author Contributions: Conceptualization, V.G., J.G. and E.K.; methodology, S.C., M.A., C.C. and D.L.; software, V.G.; validation, V.G., J.G., E.K., C.C., L.H., N.K. and D.L.; formal analysis, V.G. and S.C.; investigation, V.G., E.K., L.H., C.C and S.C.; resources, V.G., E.K., J.G., C.C., S.C., L.H., D.L. and N.K.; data curation, V.G. and E.K.; writing—original draft preparation, V.G., S.C, E.K., C.C. and J.G.; writing—review and editing, V.G., L.H. and N.K.; visualization, V.G., S.C., D.L., C.C., J.G., L.H., N.K. and E.K.; supervision, E.K., J.G, C.C, D.L., L.H. and N.K.; project administration, V.G. and E.K. All authors have read and agreed to the published version of the manuscript.

Funding: This research received no external funding.

Institutional Review Board Statement: The protocol was approved by the Ethics Committee at the University of St. Joseph, Beirut, Lebanon (USJ-2020-10, 14-1-2020).

Informed Consent Statement: Informed consent was obtained from all subjects involved in the study. Written informed consent has been obtained from the patient to publish this paper.

Data Availability Statement: The data that support the findings of this study are available from the corresponding author upon reasonable request.

Acknowledgments: Authors would like to acknowledge the Saint-Joseph University of Beirut, Lebanon. The authors would also thank University of Sorbonne Paris Nord and the University of Strasbourg for supporting this research.

Conflicts of Interest: The authors declare no conflict of interest.

References

1. Pickett, K.L.; Sadowsky, P.L.; Jacobson, A.; Lacefield, W. Orthodontic In Vivo Bond Strength: Comparison with In Vitro Results. *Angle Orthod.* **2001**, *71*, 141–148. [CrossRef] [PubMed]
2. Pinho, M.; Manso, M.C.; Almeida, R.F.; Martin, C.; Carvalho, Ó.; Henriques, B.; Silva, F.; Pinhão Ferreira, A.; Souza, J.C.M. Bond Strength of Metallic or Ceramic Orthodontic Brackets to Enamel, Acrylic, or Porcelain Surfaces. *Materials* **2020**, *13*, 5197. [CrossRef] [PubMed]
3. Grazioli, G.; Hardan, L.; Bourgi, R.; Nakanishi, L.; Amm, E.; Zarow, M.; Jakubowicz, N.; Proc, P.; Cuevas-Suárez, C.E.; Lukomska-Szymanska, M. Residual Adhesive Removal Methods for Rebonding of Debonded Orthodontic Metal Brackets: Systematic Review and Meta-Analysis. *Materials* **2021**, *14*, 6120. [CrossRef]
4. Buonocore, M.G. A Simple Method of Increasing the Adhesion of Acrylic Filling Materials to Enamel Surfaces. *J. Dent. Res.* **1955**, *34*, 849–853. [CrossRef]
5. Sharma, S.; Tandon, P.; Nagar, A.; Singh, G.P.; Singh, A.; Chugh, V.K. A Comparison of Shear Bond Strength of Orthodontic Brackets Bonded with Four Different Orthodontic Adhesives. *J. Orthod. Sci.* **2014**, *3*, 29–33. [CrossRef] [PubMed]
6. Hardan, L.; Bourgi, R.; Kharouf, N.; Mancino, D.; Zarow, M.; Jakubowicz, N.; Haikel, Y.; Cuevas-Suárez, C.E. Bond Strength of Universal Adhesives to Dentin: A Systematic Review and Meta-Analysis. *Polymers* **2021**, *13*, 814. [CrossRef]
7. Flores, A.R.; Sáez, E.G.; Barceló, F. Metallic Bracket to Enamel Bonding with a Photopolymerizable Resin-Reinforced Glass Ionomer. *Am. J. Orthod. Dentofac. Orthop.* **1999**, *116*, 514–517. [CrossRef]
8. Mickenautsch, S.; Yengopal, V.; Banerjee, A. Retention of Orthodontic Brackets Bonded with Resin-Modified GIC versus Composite Resin Adhesives–a Quantitative Systematic Review of Clinical Trials. *Clin. Oral Investig.* **2012**, *16*, 1–14. [CrossRef] [PubMed]
9. Summers, A.; Kao, E.; Gilmore, J.; Gunel, E.; Ngan, P. Comparison of Bond Strength between a Conventional Resin Adhesive and a Resin-Modified Glass Ionomer Adhesive: An in Vitro and in Vivo Study. *Am. J. Orthod. Dentofac. Orthop.* **2004**, *126*, 200–206; quiz 254–255. [CrossRef]
10. Ghoubril, V.; Ghoubril, J.; Khoury, E. A Comparison between RMGIC and Composite with Acid-Etch Preparation or Hypochlorite on the Adhesion of a Premolar Metal Bracket by Testing SBS and ARI: In Vitro Study. *Int. Orthod.* **2020**, *18*, 127–136. [CrossRef] [PubMed]
11. Sofan, E.; Sofan, A.; Palaia, G.; Tenore, G.; Romeo, U.; Migliau, G. Classification Review of Dental Adhesive Systems: From the IV Generation to the Universal Type. *Ann. Stomatol.* **2017**, *8*, 1–17. [CrossRef]

12. Blomlöf, J.; Lindskog, S. Periodontal Tissue-Vitality after Different Etching Modalities. *J. Clin. Periodontol.* **1995**, *22*, 464–468. [PubMed]
13. Blomlöf, J.; Jansson, L.; Blomlöf, L.; Lindskog, S. Long-Time Etching at Low PH Jeopardizes Periodontal Healing. *J. Clin. Periodontol.* **1995**, *22*, 459–463. [CrossRef]
14. Akman, A.C.; Demiralp, B.; Güncü, G.N.; Kiremitçi, A.; Sengün, D. Necrosis of Gingiva and Alveolar Bone Caused by Acid Etching and Its Treatment with Subepithelial Connective Tissue Graft. *J. Can. Dent. Assoc.* **2005**, *71*, 477–479.
15. Jurado, C.A.; Fischer, N.G.; Sayed, M.E.; Villalobos-Tinoco, J.; Tsujimoto, A. Rubber Dam Isolation for Bonding Ceramic Veneers: A Five-Year Post-Insertion Clinical Report. *Cureus* **2021**, *13*, e20748. [CrossRef] [PubMed]
16. Hu, H.; Li, C.; Li, F.; Chen, J.; Sun, J.; Zou, S.; Sandham, A.; Xu, Q.; Riley, P.; Ye, Q. Enamel Etching for Bonding Fixed Orthodontic Braces. *Cochrane Database Syst. Rev.* **2013**, *2013*, CD005516. [CrossRef] [PubMed]
17. Kim, D.; Kwak, J.; Jo, R.; Jung, D.; Youn, D.; Oh, N.; Jang, J. Effects of Dental Acid Etchants in Oral Epithelial Cells. *Oral Biol. Res.* **2019**, *43*, 299–305. [CrossRef]
18. Terhune, W.F.; Sydiskis, R.J.; Davidson, W.M. In Vitro Cytotoxicity of Orthodontic Bonding Materials. *Am. J. Orthod.* **1983**, *83*, 501–506. [CrossRef]
19. Park, J.-H.; Shin, H.-J.; Park, S.-H.; Kim, J.-W.; Cho, K.-M. Iatrogenic chemical burn on facial skin by 37% phosphoric acid etchant. *J. Korean Acad. Conserv. Dent.* **2009**, *34*, 38–41. [CrossRef]
20. Nakajima, A.; Kurihara, H.; Yagita, H.; Okumura, K.; Nakano, H. Mitochondrial Extrusion through the Cytoplasmic Vacuoles during Cell Death. *J. Biol. Chem.* **2008**, *283*, 24128–24135. [CrossRef] [PubMed]
21. Cummings, B.S.; Wills, L.P.; Schnellmann, R.G. Measurement of Cell Death in Mammalian Cells. *Curr. Protoc. Pharm.* **2004**, *56*, 12.8.1–12.8.24. [CrossRef] [PubMed]
22. Riss, T.L.; Moravec, R.A.; Niles, A.L.; Duellman, S.; Benink, H.A.; Worzella, T.J.; Minor, L. Cell Viability Assays. In *Assay Guidance Manual*; Markossian, S., Grossman, A., Brimacombe, K., Arkin, M., Auld, D., Austin, C., Baell, J., Chung, T.D.Y., Coussens, N.P., Dahlin, J.L., et al., Eds.; Eli Lilly & Company and the National Center for Advancing Translational Sciences: Bethesda, MD, USA, 2004.
23. Fröb, L.; Rüttermann, S.; Romanos, G.E.; Herrmann, E.; Gerhardt-Szép, S. Cytotoxicity of Self-Etch Versus Etch-and-Rinse Dentin Adhesives: A Screening Study. *Materials* **2020**, *13*, 452. [CrossRef]
24. Pupo, Y.M.; Bernardo, C.F.d.F.; de Souza, F.F.d.F.A.; Michél, M.D.; Ribeiro, C.N.d.M.; Germano, S.; Maluf, D.F. Cytotoxicity of Etch-and-Rinse, Self-Etch, and Universal Dental Adhesive Systems in Fibroblast Cell Line 3T3. *Scanning* **2017**, *2017*, 9650420. [CrossRef] [PubMed]
25. Ozer, F.; Blatz, M.B. Self-Etch and Etch-and-Rinse Adhesive Systems in Clinical Dentistry. *Compend. Contin. Educ. Dent.* **2013**, *34*, 12–14, 16, 18; quiz 20, 30. [PubMed]
26. Cecchin, D.; Farina, A.P.; Vidal, C.M.; Bedran-Russo, A.K. A Novel Enamel and Dentin Etching Protocol Using α-Hydroxy Glycolic Acid: Surface Property, Etching Pattern, and Bond Strength Studies. *Oper. Dent.* **2018**, *43*, 101–110. [CrossRef] [PubMed]

Journal of
Functional Biomaterials

MDPI

Article

A New Mass Spectroscopy-Based Method for Assessing the Periodontal–Endodontic Interface after Intracanal Placement of Biomaterials In Vitro

Andreas Braun *, Michael Berthold, Patricia Buttler, Joanna Glock and Johannes-Simon Wenzler

Department of Operative Dentistry, Periodontology and Preventive Dentistry, RWTH Aachen University Hospital, Pauwelsstrasse 30, 52074 Aachen, Germany
* Correspondence: anbraun@ukaachen.de; Tel.: +49-(0)241-80-88110; Fax: +49-(0)241-80-82468

Abstract: Optimizing the interface between biomaterials and dental hard tissues can prevent leakage of bacteria or inflammatory mediators into periapical tissues and thus avoid alveolar bone inflammation. In this study, an analysis system for testing the periodontal–endodontic interface using gas leakage and subsequent mass spectrometry was developed and validated using the roots of 15 single-rooted teeth in four groups: (I) roots without root canal filling, (II) roots with an inserted gutta-percha post without sealer, (III) roots with gutta-percha post and sealer, (IV) roots filled with sealer only, and (V) adhesively covered roots. Helium was used as the test gas, and its leakage rate was found by measuring the rising ion current using mass spectrometry. This system made it possible to differentiate between the leakage rates of tooth specimens with different fillings. Roots without filling showed the highest leakage values ($p < 0.05$). Specimens with a gutta-percha post without sealer showed statistically significantly higher leakage values than groups with a filling of gutta-percha and sealer or sealer alone ($p < 0.05$). This study shows that a standardized analysis system can be developed for periodontal–endodontic interfaces to prevent biomaterials and tissue degradation products from affecting the surrounding alveolar bone tissue.

Keywords: periodontal–endodontic interface; endodontics; permeability; interface; leakage; high vacuum; mass spectroscopy

1. Introduction

Parameters such as accuracy of fit and interactions with the existing residual tissue play decisive roles in the long-term stability and functionality of incorporated tissue replacement, and in the functioning of the hybrid structure of biological tissue and biocompatible material that has been created. The biomaterial used must be tolerated by the immune system and should initiate reparative or regenerative healing mechanisms [1,2]. If this does not happen, septic and/or aseptic disorders of healing can be expected. The unavoidable interface areas between the biological tissue and the material are a particular weak point in relation to potential healing disorders, in addition to immune defense reactions to the material used [3]. In addition, if there is a pre-existing condition that impairs blood circulation in the wound area, the prognosis for the overall therapy may be limited [4]. In this context, analysis of possible interfaces around integrated biomaterials is particularly important for patients with metabolic diseases, such as diabetes mellitus, and vascular diseases, in order to minimize disturbances in the healing process.

In endodontology and periodontology, new biocompatible materials need to be developed in such a way that they establish an adequate bond between the biomaterials used and endogenous tissue. Insufficient periodontal interfaces might otherwise cause the penetration of bacteria, necrotic tissue remnants, or inflammatory mediators that remain in the dentinal tubules in periapical tissues, leading to alveolar bone inflammation [5].

Bioactive dental glass–ceramics are suggested for dentin hypersensitivity treatment, implant coating, bone regeneration, and periodontal therapy, as they show bone-bonding ability and stimulate positive biological reactions at the material/tissue interface [6]. The possibility of a bond between bone and biomaterials is not seen as limited to bioactive materials; however, the surface topography is considered to be an essential factor for bond formation [7]. The methods used to analyze this type of interface need to respond to newly developed materials and support them with new investigation methods. Due to advances in materials development and processing, periodontal–endodontic leakage cannot be sufficiently assessed using a simple probe and now has to be assessed using more specific methods such as confocal laser scanning microscopy or microbial penetration tests [8,9]. Dye penetration tests or scanning electron microscopy, sometimes employing a replica technique, are usually used to obtain evaluable samples [10]. Analysis of passive dye diffusion is a simple technique, but it is not necessarily reliable, since influences from capillary forces, air inclusions, or dye properties such as particle size can affect the method. Other methods such as the fluid filtration method [11] and capillary flow porometry [12] use the flow of fluids through the investigated interfaces between biomaterials and dental hard substances to assess the permeability of the interface.

With the development of new materials and biocompatible endodontic sealers [13], and the continuous pursuit of better biocompatibility and durability, the requirements for adequate bonding between these materials and endogenous tissue are becoming more stringent. In order to obtain even more accurate methods for assessing the interface between biomaterials and dental hard substances, attempts have already been made to assess the permeability of an interface using leakage-induced pressure differences between the start and the end of a filled tooth root [14]. However, to the best of the authors' knowledge, the use of a test gas, the leakage of which is quantified using mass spectrometry after passage of a test specimen, has not yet been investigated.

The objective of this study was therefore to develop an analysis system for assessing interfaces using gas leakage and subsequent mass spectrometry, testing the hypothesis that it is possible to assess the periodontal–endodontic interface with this type of system. The test was conducted in a high vacuum to allow for the assessment of even the smallest particles. High-vacuum testing was also expected to improve the quantitative evaluation, due to standardized measurement parameters. The validity of the system was then investigated by examining differently filled root canal systems, testing whether the new test system is able to detect differences between the study groups. It is assumed that if a mass spectrometry-based analysis method is positively validated, it would be possible in further studies to not only conduct measurements of leakage rates at the periodontal–endodontic interface, but also to analyze selected tissue and microbial degradation products for their potential pathological impact on the periradicular tissues.

2. Materials and Methods

2.1. Measurement Setup

The test setup consists of a mass spectrometer (PrismaPro QMG 250 F1ö Pfeiffer Vacuum, Wetzlar, Germany), which analyzes the test gas and displays the results via software (Figure 1). The spectrometer is designed as a quadrupole mass spectrometry system for qualitative and quantitative gas analysis in the high and ultrahigh vacuum range. It is also designed for detecting leaks and measuring trace residues. The device is mounted via a crosspiece at an angle of 90° in the gas flow between the suctioning turbomolecular high-vacuum pump (HiPace 60 P with TC 110; Pfeiffer Vacuum, Wetzlar, Germany) with a maximum volume flow of 5×10^{-4} hPa·l·s^{-1}, and a control valve (EVR 116 gas control valve; Pfeiffer Vacuum, Wetzlar, Germany) that delivers the 99.999% high-purity test gas (Helium 5.0; Linde, Pullach, Germany). Opposite the mass spectrometer, a Pirani gauge (PKR 360; Pfeiffer Vacuum, Wetzlar) is installed to monitor the pre-vacuum pressure required for operating the spectrometer. In the event of a gas inrush, the spectrometer is shut down to prevent damage.

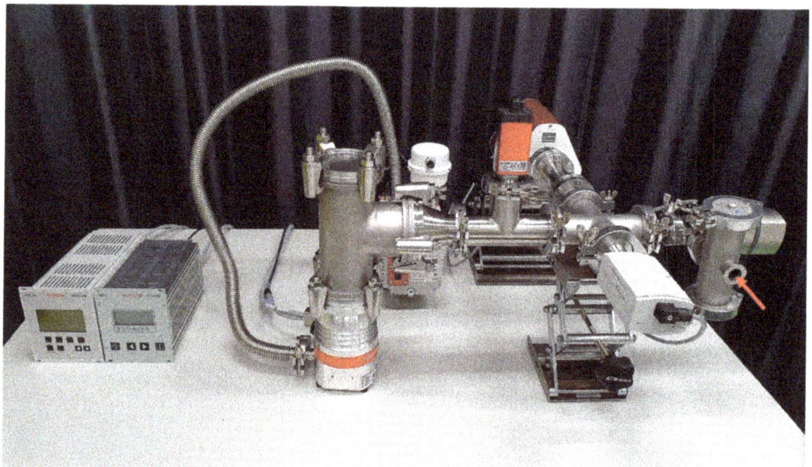

Figure 1. Test system for measuring the leakage of helium gas through the periodontal–endodontic interface. Red arrow: interface for the sample material.

The backing vacuum of 2×10^{-2} hPa required to operate the turbomolecular pump is generated by a two-stage rotary vane vacuum pump (Duo 3M; Pfeiffer Vacuum, Wetzlar, Germany). The backing and high-vacuum pumps are connected via a metal shaft hose (ISO-KF; Pfeiffer Vacuum AG, Wetzlar). The control valve for the gas inlet into the measurement setup is controlled by a control unit (RVC 300; Pfeiffer Vacuum, Wetzlar) which, in conjunction with a Bayard–Alpert type ionization vacuum gauge, forms a control loop and ensures the constant measurement of the volume flow. The measuring principle is independent of gas type, so that incorrect measurements cannot occur when different measuring, test, or purge gases are used. The measuring tube is flanged directly into the volume flow of the measurement setup by means of a T-piece between the crosspiece of the mass spectrometer and the metering valve. A measuring chamber or sample holder can be attached to the gas inlet of the gas control valve by means of an ISO-KF flange.

2.2. Sample Preparation

This study included 15 freshly extracted single-rooted human teeth from different patients. Immediately after extraction, all teeth were stored in 0.9% isotonic NaCl solution with 0.001% sodium azide. This study was conducted in full accordance with established ethical principles (World Medical Association Declaration of Helsinki, version VI, 2002). All of the patients were informed that their teeth were to be used in an in vitro research project.

The roots of the teeth were separated horizontally in the apical region with a torpedo-shaped dental diamond burr (Brasseler, Lemgo, Germany) at 40,000 rpm, resulting in a residual root length of 4 mm and exposing a root canal in all cases. The root canals were checked for patency with an ISO size 10 file (K-file; VDW, Munich, Germany), and the intended root canal preparation length of 3 mm from the horizontal section level was checked using a radiograph with an ISO size 15 file inserted. Subsequently, the root canals were manually prepared up to ISO size 35 employing sodium hypochlorite 3% and ethylenediaminetetraacetic acid 17% rinsing solutions, and a check for patency was also performed again after preparation. The root specimens prepared in this way were then divided into five study groups:

I. Roots without filling of the prepared root canal (positive control).
II. Roots with an inserted gutta-percha post (ISO standardized gutta-percha; VDW, Munich, Germany) corresponding to the preparation size (without sealer).
III. Roots with an inserted gutta-percha post corresponding to the preparation size with sealer AH Plus (Dentsply Sirona, Bensheim, Germany).

IV. Roots without an inserted gutta-percha post, filled with sealer only in the prepared area.

V. Roots covered with a high-vacuum adhesive (IB-UHK 2020; iBEGO, Bochum, Germany) (negative control).

The same roots were used in the same sequence in each group. Any filling material present was completely removed from the root canal at the beginning of the experimental procedure, and the canal system was checked again for patency in each case. A representative image of a root specimen is shown in Figure 2.

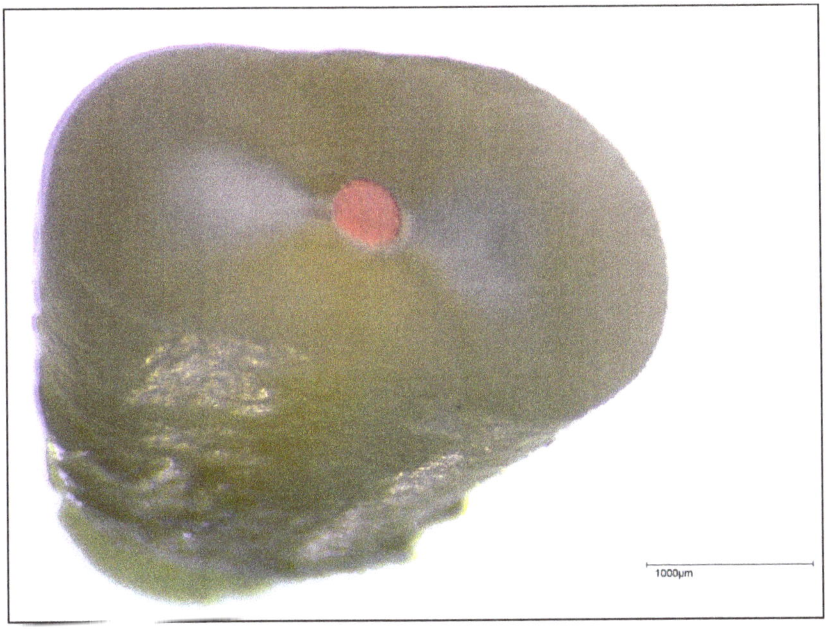

Figure 2. Representative image of a root specimen used for leakage measurements.

For measurement of the processed specimens, sample holding devices were 3D printed in the form of an axially perforated cylinder and a trough rounded out toward the specimen. On the side facing away from the specimen, there was a tube for connecting the specimen holding device to an adapter flange for connection to the measurement setup. The specimens were fixed in the trough of the specimen holding device using a high-vacuum adhesive (IB-UHK 2020) and stored in an oven at 35 °C for 24 h to cure the adhesive. To avoid contamination of the spectrometer and vacuum system with water vapor after 24 h of water storage at 37 °C, the specimens were dried and stored in individually sealed containers with silica gel beads until measurement. For the individual measurements, the sample was placed on the adapter flange with the metering valve closed. The sample was then exposed to the helium test gas for 15 s with the metering valve open and set to a flow rate of 60 cm^3/h. With a single measurement time of 32 ms, 469 measurements were performed within the respective measurement interval of 15 s. During the entire measurement cycle, beginning with the opening of the gas metering valve and ending with its closing, the helium spectrum after the passage of the sample was evaluated with a mass spectrometer and quantitatively displayed as ion current [A].

2.3. Statistical Analysis

A power analysis was performed prior to this study. The Cohen effect size was set to 0.8 [15]. For an alpha error of 0.05 and a power of 0.8, a sample size of at least 10 specimens in each group was calculated. The normal distribution of the values was assessed using the Shapiro–Wilk test. Since not all data were normally distributed, values were analyzed using a nonparametric test for dependent samples (Friedmann) and Wilcoxon pairwise comparisons. Sequentially rejective Bonferroni correction of the critical p value was used when multiple statistical tests were performed simultaneously on a single data set. Differences were considered statistically significant at $p < 0.05$. Box plot diagrams show the median, the first and third quartiles, and the minimum and maximum values (whiskers). Values of more than 1.5–3 times the interquartile range were specified as outliers and marked as data points. Values more than three times the interquartile range were specified as far outliers and marked as asterisks.

3. Results

Leakage of the test gas was detected in all of the samples and measured as ion current. Statistically significantly different leakage rates were observed in the study groups investigated ($p < 0.05$) (Figure 3). Study group I (positive control) comprised the root apices of natural teeth, in which the root canal was mechanically prepared up to the physiological foramen. Since all of the preparations were tested for patency, it was ensured that an unobstructed root foramen allowed a continuous connection out of the root canal even after the area was prepared. The highest values for gas leakage among all the study groups were found in this group, with a median value of 1.4×10^{-8} [A] (min. 2.6×10^{-9}, max. 6.6×10^{-8}, interquartile range 9.4×10^{-9}) ($p < 0.05$).

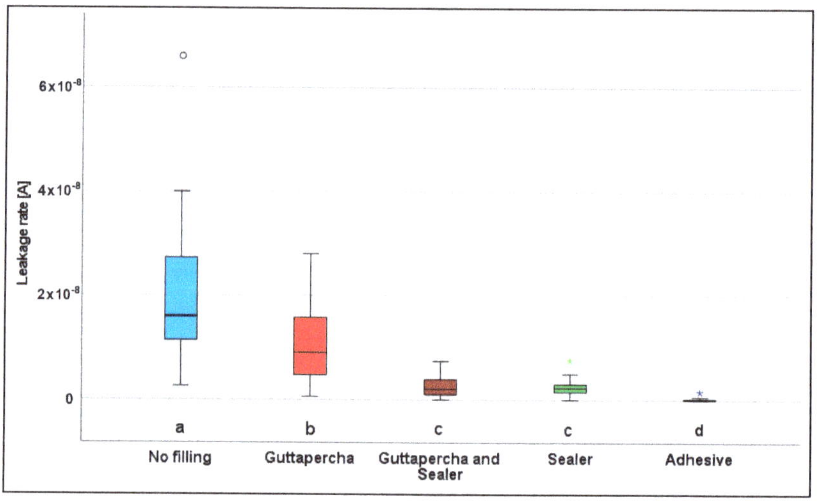

Figure 3. Box plot diagram for the ion current in the different study groups. Values of more than 1.5–3 times the interquartile range were specified as outliers and marked as data points "°". Values more than three times the interquartile range were specified as far outliers and marked as asterisks "*". Different indices (a–d) indicate groups with statistically significant differences ($p < 0.05$).

Study group II differed in that the prepared canal was obturated up to the physiological foramen with a gutta-percha post matching the preparation size without the use of a sealer. A statistically significant reduction in the leakage rate was observed in comparison with the first group, with a median value of 1.1×10^{-8} (min. 6.0×10^{-10}, max. 2.8×10^{-8}, interquartile range 1.1×10^{-8}) ($p < 0.05$).

In study group III, the use of a sealer between the gutta-percha post and the canal wall was intended to supplement possible inaccuracies. Statistically significantly lower leakage rates were detected in this group in comparison with the first two groups, with a median value of $1.9 \times 10{-}9$ (min. 4.0×10^{-11}, max. 7.5×10^{-9}, interquartile range 2.6×10^{-9}) ($p < 0.05$).

In test group IV, the filling of the root canal was performed with sealer paste alone. In comparison with groups I and II, statistically significantly lower leakage values were also observed here, with a median value of 2.0×10^{-9} (min. 2.0×10^{-11}, max. 7.6×10^{-9}, interquartile range 1.6×10^{-9}) ($p < 0.05$). However, comparison with study group III did not show any statistically significant differences ($p > 0.05$).

In the negative controls (group V), the horizontal cutting surface of the root was covered with a high-vacuum adhesive. The statistically significantly lowest gas leakage was observed in all cases in comparison with all of the other groups, with a median value of 3.1×10^{-11} (min. 8.7×10^{-12}, max. 1.6×10^{-9}, interquartile range 2.9×10^{-10}) ($p < 0.05$) (Table 1).

Table 1. Maximum ion current [A] in each study group.

	No Filling	Gutta-Percha	Gutta-Percha and Sealer	Sealer	Adhesive
Mean	2.2×10^{-8}	1.1×10^{-8}	2.6×10^{-9}	2.6×10^{-9}	2.3×10^{-10}
Standard deviation	1.6×10^{-8}	8.1×10^{-9}	2.1×10^{-9}	1.9×10^{-9}	4.1×10^{-10}
Median	1.4×10^{-8}	1.1×10^{-8}	1.9×10^{-9}	2.0×10^{-9}	3.1×10^{-11}
Minimum	2.6×10^{-9}	6.0×10^{-10}	4.0×10^{-11}	2.0×10^{-11}	8.7×10^{-12}
Maximum	6.6×10^{-8}	2.8×10^{-8}	7.5×10^{-9}	7.6×10^{-9}	1.6×10^{-9}
Interquartile range	9.4×10^{-9}	1.1×10^{-8}	2.6×10^{-9}	1.6×10^{-9}	2.9×10^{-10}
Number	15	15	15	15	15

4. Discussion

This study investigated the development of an analysis system for testing periodontal–endodontic interfaces in a high-vacuum setting in relation to the permeation of various biomaterials into periapical tissues. It shows promising results.

Until now, leakage testing of root canal fillings or their materials could only be performed approximately using dye, bacteria, or glucose penetration tests; tests with radioactive markers; or by checking the integrity of margins using scanning electron microscopy [16]. However, these methods almost always showed very heterogeneous results, as they are too sensitive and prone to errors, among other things, due to the pH dependence of the color solution, the size of the molecule, or a lack of quantification options. Additionally, due to the semiquantitative evaluation of the results, there is often a certain lack of clarity, and therefore the corresponding significance is of extremely low clinical relevance [17,18]. In order to take these factors into account or neutralize them, this study recorded even the smallest particles purely quantitatively in a high-vacuum setting, with an option for further mass spectrometry analysis.

The dye penetration method (DPM) is the one most widely used. In this technique, the penetration depth of the dye is considered to correlate with the leakiness of the root filling [18]. It has advantages in terms of sensitivity, for example, but it is also not an accurate examination method. Another approach is the fluid filtration method (FFM), which is based on a liquid being passed through the materials to be examined, such as the sealer, resulting in cavities between the sealer and the dentin walls, and the sealer and the gutta-percha. This method was proposed many years ago as a new reference technique for leakage investigations [14,18]. Other techniques, such as capillary flow porometry (CFP), are independent of the wetting properties of the biomaterials at the interface being investigated [12]. The approach used in the present study tries to adopt the positive characteristics of the older methods, while at the same time being able to record the results and confirm their statistical significance. A similar approach, albeit on a different scale, but also using the gas permeability method (GPM), demonstrated the feasibility of this methodology to some extent [14]. However, in contrast to that study, which used nitrogen, the present study used helium as the test gas. The decision to use helium was based on the fact that it is the most commonly used gas for leak detection, and is also inert, nontoxic, and has a high diffusion capacity. As with the findings of this study, in the study previously mentioned, it became apparent how important it is to develop a method that is free of hydrophobic/hydrophilic interactions, for example, when considering biomaterials, interfaces, and leakages; in many of the methods used so far, the results are often contradictory and not comparable [17–20]. The lack of standardization seems to be just as decisive here as the variance, even within the different methods—very few results were really reproducible [21]. An interesting approach, which can perhaps be seen as a precursor and therefore not yet fully developed, investigated the possibility of leak testing using compressed air (CA), in which pore diameters of only 0.12 μm can be detected at an air pressure of 25 atm (physical atmosphere) [22]. Admittedly, the pressure values used are scarcely comparable. For compressed air, 25 atm is the equivalent of 25,331 hPa, which of course seems too high against the background of the CFP method, with up to 13,789 hPa; the fluid filtration method, with up to 1200 hPa; and the GP method, with approximately 990 hPa. However, it should be noted that the compressed air test, unlike the GP test, for example, is not an inherently closed system. The system used in the present study is also closed, with a maximum pressure difference of 900–1020 hPa between the vacuum and the environment. In comparison with the other methods, in this experimental setup it cannot be expected that a pressure constant that is too high could dissolve the materials. To confirm this again, it should be mentioned that the sample gas in the test chamber is not pressed into the samples, but gently flows around them, even with a helium supply of 60 cm^3/h. It is diffused into the sample due to the negative pressure and is thus not a significant factor for leakage in itself. For comparison, Romieu et al. set values of 3.5×10^7 mol/s or 0.5 cm^3/h for the gas flow, but with an experimental setup that differs from ours [14].

As in the study by Romieu et al. [14], the present results also showed that varying leakage rates in the different study groups could be successfully identified. Even assuming that absorption of the test gas on the surfaces or into the depths of the samples cannot be excluded, an intraexperimental comparison of the study groups is possible, since such absorption can be assumed to be similar in all groups. Comparisons between study group I (open root canal; positive control group), study group II (obturation with gutta-percha), study group III (obturation with gutta-percha and biomaterial-based sealer), study group IV (obturation with a biomaterial-based sealer only), and study group V (adhesively sealed sample; negative control group) in most cases showed statistically significant differences. It should be added that the clear statistical significance of the positive and negative control tests confirmed that the experimental setup works and can be used without restrictions.

The different filling techniques mentioned above were also an important factor in this study, since different materials [23,24] and the layer thickness used in the techniques [25–27] can of course have a considerable influence on seal tightness. This was also confirmed with the method. The results show clear statistically significant differences between the control group (I) and the definitive filling groups (III and IV). However, there is still a need for further research, e.g., studies testing the seal tightness between lateral condensation, the single cone technique, and warm filling techniques or other biomaterials.

The length, root canal preparation, and irrigation of the sample must of course also be critically considered. The samples were all 4 mm in length and were repeatedly prepared (filling retreatment) as connected samples within the group. The length appears to be very short in comparison with other FF studies. Many studies, e.g., those based on the dye penetration method, use lengths of 10–15 mm [25,28]. However, this sample length was deliberately chosen for the feasibility of this study, in order to avoid influencing factors such as possible curved canals or furcations, the formation of air pockets in the root fillings, and to obtain meaningful and comparable results. Connected samples also meant that they had to be cleaned after each measurement, i.e., the filling materials had to be carefully revised. Of course, unintentional residues of materials in the samples could lead to distortions in the evaluation [29]. This can certainly also be interpreted in the results with regard to sample groups III and IV. It seems interesting that there was no statistically significant difference between the two groups and that there was a smaller interquartile range in group IV. Here, due to the missing range, one could certainly conclude that there were residues from previous filling materials from the previous run. The intention was to counteract this possibility as much as possible by using a small sample length. Visually and tactilely, an effort was made to ensure the removal of the previous biomaterial during the filling material retreatment; however, total removal of the biomaterial could not be guaranteed.

Another point that should also be considered here is the influence of the rinsing solutions used in the treatment on the adhesive strength of the biomaterials in the root canal. We used the Aachen rinsing protocol for chemomechanical preparation, which involves rinsing with a conventional 17% EDTA rinsing solution and 3% sodium hypochlorite rinsing solution. Both rinsing solutions are very well established and have been widely studied in the literature, including in connection with the filling materials used in this study, and in relation to bond strength [30]. Due to the large number of published studies on these rinsing solutions, including combinations with different types of bioactive sealants and their bonds, closer attention to this aspect seems unnecessary.

The present feasibility study provides a good insight into the analysis of interface leakage in the high-vacuum system. However, further experiments will be needed to allow more precise conclusions to be drawn. It would also be interesting to assess the extent to which different biomaterial-based (root) filling materials can prevent the leakage of sulfur compounds formed by persistent micro-organisms at the periodontal–endodontic interface, which have been described as potentially harmful [31]. There are thus exciting prospects for further research, which should definitively deal with the above-mentioned potential infiltration of toxins, using different root filling techniques and bond strengths, for example.

5. Conclusions

This study describes the development of a standardized test setup for assessing periodontal–endodontic interfaces in order to evaluate and optimize hybrid structures between root dentin and endodontic sealing materials. Compared to current analysis systems, it is possible not only to record leakage rates in general, but also to quantitatively measure the smallest particles, with the option of further mass spectrometric analysis. Verifying the sealing of a root canal can prevent the surrounding alveolar bone tissue from being affected by biomaterials and tissue degradation products. Determining the sealing properties of restorative biomaterials and material combinations should allow for better assessment of the prognosis for avoiding apical inflammatory processes and achieving apical healing.

Author Contributions: Conceptualization, A.B., M.B. and J.-S.W.; methodology, M.B.; formal analysis, A.B., M.B., P.B., J.G. and J.-S.W.; investigation, M.B., P.B. and J.G.; resources, A.B.; data curation, A.B.; writing—original draft preparation, A.B., M.B. and J.-S.W.; writing—review and editing, A.B., M.B., J.G., P.B. and J.-S.W.; supervision, A.B. and J.-S.W.; project administration, A.B. and J.-S.W.; funding acquisition, A.B. and J.-S.W. All authors have read and agreed to the published version of the manuscript.

Funding: This study was supported by a grant from the Interdisciplinary Center for Clinical Research in the Faculty of Medicine at RWTH Aachen University (OC1-12).

Institutional Review Board Statement: No. EK 141/21, Ethics Committee of the Medical Faculty of the University Hospital RWTH Aachen.

Informed Consent Statement: This study was conducted in full accordance with established ethical principles (World Medical Association Declaration of Helsinki, version VI, 2002). All of the patients were informed that their teeth were to be used in an in vitro research project.

Data Availability Statement: Data supporting the results reported here can be requested by e-mail.

Acknowledgments: The authors would like to thank the entire team of the scientific laboratory in the Department of Operative Dentistry, Periodontology and Preventive Dentistry at RWTH Aachen University Hospital for their support.

Conflicts of Interest: The authors declare that they had no conflict of interest.

References

1. Marin, E.; Boschetto, F.; Pezzotti, G. Biomaterials and biocompatibility: An historical overview. *J. Biomed. Mater. Res.* **2020**, *108*, 1617–1633. [CrossRef] [PubMed]
2. Murabayashi, S.; Nose, Y. Biocompatibility: Bioengineering aspects. *Biomed. Mater. Eng.* **2013**, *23*, 129–142. [CrossRef]
3. Raut, H.K.; Das, R.; Liu, Z.; Liu, X.; Ramakrishna, S. Biocompatibility of biomaterials for tissue regeneration or replacement. *Biotechnol. J.* **2020**, *15*, 2000160. [CrossRef] [PubMed]
4. Gadicherla, S.; Smriti, K.; Roy, S.; Pentapati, K.-C.; Rajan, J.; Walia, A. Comparison of extraction socket healing in non-diabetic, prediabetic, and type 2 diabetic patients. *Clin. Cosmet. Investig. Dent.* **2020**, *12*, 291–296. [CrossRef] [PubMed]
5. Holland, R.; Gomes Filho, J.E.; Cintra, L.T.A.; Queiroz, Í.O.D.A.; Estrela, C. Factors affecting the periapical healing process of endodontically treated teeth. *J. Appl. Oral Sci.* **2017**, *25*, 465–476. [CrossRef]
6. Montazerian, M.; Zanotto, E.D. Bioactive and inert dental glass-ceramics. *J. Biomed. Mater. Res. A* **2017**, *105*, 619–639. [CrossRef]
7. Davies, J.E. Bone bonding at natural and biomaterial surfaces. *Biomaterials* **2007**, *28*, 5058–5067. [CrossRef]
8. Hwang, J.H.; Chung, J.; Na, H.-S.; Park, E.; Kwak, S.; Kim, H.-C. Comparison of bacterial leakage resistance of various root canal filling materials and methods: Confocal laser-scanning microscope study: Bacterial leakage resistance of canal filling materials. *Scanning* **2015**, *37*, 422–428. [CrossRef]
9. Savadkouhi, S.; Bakhtiar, H.; Ardestani, S. In vitro and ex vivo microbial leakage assessment in endodontics: A literature review. *J. Int. Soc. Prevent. Communit. Dent.* **2016**, *6*, 509. [CrossRef]
10. Çiftçi, A.; Vardarlı, D.A.; Sönmez, I.Ş. Coronal microleakage of four endodontic temporary restorative materials: An in vitro study. *Oral Surg. Oral Med. Oral Pathol. Oral Radiol. Endod.* **2009**, *108*, e67–e70. [CrossRef]
11. Oruçoğlu, H.; Sengun, A.; Yilmaz, N. Apical leakage of resin based root canal sealers with a new computerized fluid filtration meter. *J. Endod.* **2005**, *31*, 886–890. [CrossRef]
12. De Bruyne, M.A.A.; De Bruyne, R.J.E.; De Moor, R.J.G. Capillary flow porometry to assess the seal provided by root-end filling materials in a standardized and reproducible way. *J. Endod.* **2006**, *32*, 206–209. [CrossRef]
13. Komabayashi, T.; Colmenar, D.; Cvach, N.; Bhat, A.; Primus, C.; Imai, Y. Comprehensive review of current endodontic sealers. *Dent. Mater. J.* **2020**, *39*, 703–720. [CrossRef]
14. Romieu, O.J.; Jacquot, B.; Callas-Etienne, S.; Dutilleul, P.-Y.C.; Levallois, B.; Cuisinier, F.J.G. Gas permeability: A new quantitative method to assess endodontic leakage. *Biomed. Tech.* **2008**, *53*, 181–184. [CrossRef]
15. Cohen, J. *Statistical Power Analysis for the Behavioral Sciences*, 2nd ed.; Taylor and Francis: Hoboken, NJ, USA, 2013; ISBN 978-1-134-74270-7.
16. Kqiku, L.; Miletic, I.; Gruber, H.J.; Anic, I.; Städtler, P. Dichtigkeit von Wurzelkanalfüllungen mit GuttaFlow® und Resilon™ im Vergleich zur lateralen Kondensation. *Wien. Med. Wochenschr.* **2010**, *160*, 230–234. [CrossRef]
17. Susini, G.; Pommel, L.; About, I.; Camps, J. Lack of correlation between ex vivo apical dye penetration and presence of apical radiolucencies. *Oral Surg. Oral Med. Oral Pathol. Oral Radiol. Endod.* **2006**, *102*, e19–e23. [CrossRef]
18. Wu, M.-K.; Wesselink, P.R. Endodontic leakage studies reconsidered. Part I. Methodology, application and relevance. *Int. Endod. J.* **1993**, *26*, 37–43. [CrossRef]
19. Michaïlesco, P.M.; Valcarcel, J.; Grieve, A.R.; Levallois, B.; Lerner, D. Bacterial leakage in endodontics: An improved method for quantification. *J. Endod.* **1996**, *22*, 535–539. [CrossRef]
20. Kersten, H.W.; Moorer, W.R. Particles and molecules in endodontic leakage. *Int. Endod. J.* **1989**, *22*, 118–124. [CrossRef]
21. Camps, J.; Pashley, D. Reliability of the dye penetration studies. *J. Endod.* **2003**, *29*, 592–594. [CrossRef]
22. Nielsen, T.H. Sealing Ability of chelate root filling cements: The triple CA-test using compressed air, part 3. *J. Endod.* **1980**, *6*, 835–841. [CrossRef] [PubMed]
23. Lee, K.; Williams, M.; Camps, J.; Pashley, D. Adhesion of endodontic sealers to dentin and gutta-percha. *J. Endod.* **2002**, *28*, 684–688. [CrossRef] [PubMed]
24. Saleh, I.; Ruyter, I.; Haapasalo, M.; Orstavik, D. Adhesion of endodontic sealers: Scanning electron microscopy and energy dispersive spectroscopy. *J. Endod.* **2003**, *29*, 595–601. [CrossRef] [PubMed]
25. Vo, K.; Daniel, J.; Ahn, C.; Primus, C.; Komabayashi, T. Coronal and apical leakage among five endodontic sealers. *J. Oral. Sci.* **2022**, *64*, 95–98. [CrossRef]
26. Libonati, A. Microbial leakage evaluation of warm gutta–percha techniques. *J. Biol. Regul. Homeost. Agents* **2021**, *35*, 67–75. [CrossRef]
27. Wu, M.-K.; Van Der Sluis, L.W.M.; Ardila, C.N.; Wesselink, P.R. Fluid movement along the coronal two-thirds of root fillings placed by three different gutta-percha techniques: Evaluation of three obturation techniques. *Int. Endod. J.* **2003**, *36*, 533–540. [CrossRef]

28. Lahor-Soler, E.; Miranda-Rius, J.; Brunet-Llobet, L.; Farre, M.; Pumarola, J. In vitro study of the apical microleakage with Resilon root canal filling using different final endodontic irrigants. *J. Clin. Exp. Dent.* **2015**, *7*, e212–e217. [CrossRef]
29. Rossi-Fedele, G.; Ahmed, H.M.A. Assessment of root canal filling removal effectiveness using micro-computed tomography: A systematic review. *J. Endod.* **2017**, *43*, 520–526. [CrossRef]
30. Assmann, E.; Scarparo, R.K.; Böttcher, D.E.; Grecca, F.S. Dentin bond strength of two mineral trioxide aggregate-based and one epoxy resin-based sealers. *J. Endod.* **2012**, *38*, 219–221. [CrossRef]
31. Jacobi-Gresser, E.; Schütt, S.; Huesker, K.; Von Baehr, V. Methyl mercaptan and hydrogen sulfide products stimulate proinflammatory cytokines in patients with necrotic pulp tissue and endodontically treated teeth. *J. Biol. Regul. Homeost. Agents* **2015**, *29*, 73–84.

Journal of
Functional Biomaterials

MDPI

Article

Efficacy of Endodontic Disinfection Protocols in an *E. faecalis* Biofilm Model—Using DAPI Staining and SEM

Maria Dede [1,2,*], Sabine Basche [1], Jörg Neunzehn [3], Martin Dannemann [4], Christian Hannig [1] and Marie-Theres Kühne [1,*]

1 Policlinic of Operative and Pediatric Dentistry, Medical Faculty Carl Gustav Carus, TU Dresden, 01307 Dresden, Germany
2 Department of Endodontics, School of Dentistry, National and Kapodistrian University of Athens, 11527 Athens, Greece
3 Vertriebsgesellschaft GmbH, Geistlich Biomaterials, Schneidweg 5, 76534 Baden-Baden, Germany
4 Faculty of Automotive Engineering, Institute of Energy and Transport Engineering, Westsächsische Hochschule Zwickau, 08056 Zwickau, Germany
* Correspondence: demarie@windowslive.com (M.D.); marie-theres.kuehne@uniklinikum-dresden.de (M.-T.K.); Tel.: +49-0351-458-4168 (M.D. & M.-T.K.); Fax: +49-0351-458-5381 (M.D. & M.-T.K.)

Abstract: The aim of this study was to investigate the antimicrobial efficacy of different disinfection protocols in a novel *Enterococcus faecalis* biofilm model based on a visualization method and to evaluate the potential alteration of dentinal surface. A total of 120 extracted human premolars were allocated to 6 groups with different irrigation protocols. The assessment of the effectiveness of each protocol and the alteration of dentinal surface were visualized by using SEM and fluorescence microscopy (DAPI). A dense *E. faecalis* biofilm with a penetration depth of 289 μm (medial part of the root canal) and 93 μm (apical part) validated that the biofilm model had been successfully implemented. A significant difference between the 3% NaOCl groups and all the other groups in both observed parts of the root canal ($p < 0.05$) was detected. However, the SEM analysis revealed that the dentinal surface in the 3% NaOCl groups was severely altered. The established biofilm model and the visualization method based on DAPI are appropriate for bacterial quantification and evaluation of the depth effect of different disinfection protocols in the root canal system. The combination of 3% NaOCl with 20% EDTA or MTAD with PUI allows the decontamination of deeper dentine zones within the root canal but simultaneously alters the dentinal surface.

Keywords: bacterial penetration; biofilm model; DAPI method; dentinal tubules; *Enterococcus faecalis*; root canal irrigants

1. Introduction

The goal of an endodontic treatment should be the elimination of microorganisms and the prevention of a possible reinfection. A successful root canal therapy relies on the combination of proper instrumentation, irrigation, and obturation of the root canal system. The etiology behind endodontic treatment failures is mainly a persisting infection with a biofilm structure in the root canal system [1]. Unfortunately, the root canal system with its anatomical complexity represents a challenging environment for the effective removal of bacteria and biofilm [2]. A plethora of chemical irrigants activated by different technical devices are used to eliminate residual microbes in root canals [2,3]. One of the most common techniques is passive ultrasonic irrigation (PUI) [4–6].

A microorganism that has been intimately associated with treatment failures is *Enterococcus faecalis* (*E. faecalis*) [7,8]. The reported prevalence of *E. faecalis* ranges from 24% to 77% in post-treatment root canal infections [9]. To date, it is still not possible to explain this prevalence since the origin of *E. faecalis* infections remains unknown. Since its first description in 1906, it is termed *Streptococcus faecalis* or "Streptococcus of faecal origin" as it has often been recovered from fecal matter or sewage [10]. It is both a commensal pathogen of the gastro-intestinal tract and a common nosocomial pathogen. Its transition from commensal to pathogenic is far from being completely understood. *E. faecalis* is most likely not derived from the endogenous flora or from nosocomial transmission but is instead a food-borne pathogen in root canal infections [11]. However, *E. faecalis* is considered a suitable model for studying bacterial infections in root canals [12], specifically in in vitro studies.

Extensive research has been accomplished in the field with regard to bacterial reduction of biofilm within the root canal system [13,14]. However, there are limited studies that compared the synergistic effects of passive ultrasonic irrigation with different irrigants against *E. faecalis* biofilm in the root canal system, as well as their effects on the dentinal root canal surface. The use of passive ultrasonic irrigation has been limited to endodontic irrigants, such as NaOCl and EDTA, and its use over CHX or MTAD (mixture of tetracycline, acid, and detergent), whereas the combination of different irrigants has not been studied in detail [6,15].

Microscopic techniques have been used for the evaluation of the effects of various endodontic irrigants on biofilms [16,17]. The dentinal root canal surface was evaluated by scanning electron microscopy (SEM) in this study. Fluorescence staining with DAPI—a fluorescent dye to visualize bacteria by binding to the AT-rich regions of nucleic acids of double-stranded DNA, thereby forming fluorescent units—was used to detect bacteria in the depth of dentinal tubules [18,19]. In the present study, the degradation and removal effects of different disinfection protocols were investigated using a visualization method. The aim of our study was to establish a biofilm root canal model in order to visualize and quantify bacterial colonization within the dentinal tubules after the application of different disinfection protocols and to identify the synergistic effects of the different irrigants in combination with PUI on the dentinal surface for the first time.

2. Materials and Methods

2.1. Sample Preparation

A total of 120 extracted human single-rooted premolar teeth were used in this study. The anatomic crown of each tooth was resected horizontally at 16 mm with a diamond disk (Diamond Disc 330 CA, Struers, Willich, Germany). The working length of each root canal was measured at 15 mm. The roots were prepared using the file F1 (020/06) of the rotary endodontic nickel–titanium system, ProTaper Universal (Dentsply-Maillefer, Switzerland), and were irrigated with 2 mL of NaCl (0.9%). Finally, the root canals were rinsed with 2 mL of 20% EDTA for 1 min under agitation with the ultrasonic tip IRRI S/25 mm (Satelec-VDW GmbH, Munich, Germany) for the removal of the smear layer and again with 0.9% NaCl. In order to prepare all samples for the splitting of the roots at the end of the procedure, two lines were drawn longitudinally on the buccal and lingual planes of each root. Longitudinal grooves were then carved with a diamond bur under the caution of not invading the root canal along the drawn lines (Figure 1). The complete longitudinal fracture of the roots was performed with a razorblade (Herkenrath, Solingen, Germany) and a hammer (Braun, Tuttlingen, Germany), providing two root halves from each sample.

Figure 1. (**A**) Initial decoronized root; (**B**) carving a groove with a diamond bur without invading the root canal; and (**C**) carved groove on the lingual and buccal planes of the root.

Lastly, the teeth were cleaned in an ultrasonic bath (Sonorex Digital 10P, Bandelin, Berlin, Germany) with 20% EDTA (10 min) and finally in distilled water (1 h) (Aqua destillata, Weinert Wassertechnik GmbH, Dresden, Germany). The apical foramen of each root and the lateral grooves were sealed with silicone (Provil novo, Heraeus Kulzer GmbH, Germany) to avoid bacterial leakage through the apex and the lateral canals or through the dentinal tubules during the procedure of inoculation with *E. faecalis*. Afterward, the teeth were placed in an ultrasonic bath for 10 min with tryptic soy broth (TSB—Merck, Darmstadt, Germany), followed by autoclaving (10 min, 121 °C). To check the sterility of the samples, the teeth were incubated in the TSB for three days at 37 °C to prove that no contamination, as indicated by the cloudiness of the TSB, took place. Afterward, the teeth were embedded in a 5 mL Eppendorf tube (Eppendorf AG, Hamburg, Germany) with 3% Agarose (Merck, Darmstadt, Germany). Again, each root was filled with the TSB medium using a fine 27-gauge needle (3/4 0.4 mm × 19 mm) (Transcodent, Germany) to keep the dentine moistened.

2.2. Inoculation of the Roots with E. faecalis and Incubation

The *E. faecalis* strain was obtained (clinical bacterium isolated from a patient with persistent root canal infection and approved by the ethics committee of the University of Freiburg 140/09) and cultivated in a S2 laboratory. A day prior to the inoculation procedure of the roots, a tryptic-soy-broth (TSB) bouillon, including 2 mg/mL of Streptomycin and 0.2 μg/mL of Amphotericin B, was inoculated with a single *E. faecalis* colony on an agar plate and incubated at 37 °C. Streptomycin (2 mg/mL) was added as an antibacterial agent to the TSB in order to inhibit the growth of other bacteria and Amphotericin B was added as an antimycotic agent. On the first day of the inoculation procedure, each root in the Eppendorf tube was inoculated with *E. faecalis* culture bouillon (1.6×10^8/mL) after the removal of the old TSB medium and incubated at 37 °C. The next day, the bacterial suspension within the root canal was again removed with a 27-gauge needle, and new *E. faecalis* culture bouillon was applied to the roots and incubated at 37 °C. After two days, the bacterial suspension was removed, but this time, it was not renewed; instead, fresh TSB medium was applied to the roots. Furthermore, the infected root canals underwent a renewal of the TSB medium every day for 6 weeks. The incubation time lasted 6 weeks to allow the formation of *E. faecalis* biofilm.

2.3. Application of the Disinfection Protocols

After the incubation period, the bacterial suspension was removed from all the roots, and the samples were randomly divided into five experimental groups (n = 20) and one control group (n = 20). Group 1 (CTR): control group with no disinfection protocol; Group 2 (NEC_{PUI}): NaOCl 3% + EDTA 20% + CHX 2% + NaCl 0.9% under passive ultrasonic irrigation (PUI) with Irri S 25 (Satelec-VDW GmbH, Munich, Germany); Group 3 (NE_{PUI}): NaOCl 3% + EDTA 20% under PUI with Irri S 25; Group 4 (CE_{PUI}): CHX 2% + EDTA 20% under PUI with Irri S 25; Group 5 (NM_{PUI}): NaOCl 3% + MTAD under PUI with Irri S25; and Group 6 ($NaCL_{PUI}$): NaCl 0.9% under PUI with Irri S 25 (Table 1). Before the disinfection procedure of any group (including the control group), the roots of all groups were instrumented up to file F4 (040/0.6) of the ProTaper system (Dentsply-Maillefer, Ballaigues, Switzerland) and were rinsed intermittently with 0.9% NaCl solution.

Table 1. Disinfection/irrigation protocols in each group.

Group 1 CTR	Group 2 NEC_{PUI}	Group 3 NE_{PUI}	Group 4 CE_{PUI}	Group 5 NM_{PUI}	Group 6 $NaCl_{PUI}$
Control group no disinfection protocol	NaOCl 3% + EDTA 20% + CHX 2% + NaCl 0.9% under passive ultrasonic irrigation (PUI) with Irri S 25	NaOCl 3% + EDTA 20% under PUI with Irri S 25	CHX 2% + EDTA 20% under PUI with Irri S 25	NaOCl 3% + MTAD under PUI with Irri S 25	NaCl 0.9% under PUI with Irri S 25

2.4. Preparation of the Specimens for Visualization Techniques

The splitting of the roots was performed directly after the application of each disinfection protocol, providing two halves of each sample. One-half of the root was used for the analysis by scanning electron microscopy (SEM), and the other half was used for the analysis by fluorescence microscopic staining with 4′,6-Diamindin-2-phenylindole (DAPI). The halves for DAPI staining were transferred directly into a 15 mL centrifugal tube (Sarstedt, Nürnberg, Germany) with 4% formaldehyde to fixate the bacteria. The fixation lasted 48 h at 4 °C. Subsequently, the roots were placed in Osteosoft® (Merck, Darmstadt, Germany)—for the decalcification of the dentinal roots—until the specimens were sliceable with a scalpel. Then, every root canal half was cut transversally into two pieces: the medial part and the apical part (1 mm before the apex). Shortly after, the embedding of the medial and apical root pieces in paraffin was carried out. The sectioning of the root pieces into 2 μm sections with a microtome (Leica Biosystems Nussloch GmbH, Nußloch, Germany) took place. The object carrier was silanized, and the test species were placed on glass on a histological slide. These final steps were performed for visualization with the DAPI method.

2.5. DAPI

DAPI staining (Merck, Darmstadt, Germany) was conducted as described in previous research [20,21]. DAPI (4′,6-diamidino-2-phenylindole) stains DNA unspecifically by binding to the AT-rich regions of double-stranded DNA. The following steps were used for the DAPI staining. The test species were covered with the DAPI stock solution (1.5 μL of stock solution in 500 μL of PBS (phosphate-buffered saline—Invitrogen Ltd., Bend, OR, USA)) in a dark chamber. This DAPI solution was removed after 15 min by rinsing several times with the PBS (phosphate-buffered saline, Invitrogen Ltd., Bend, OR, USA) before the samples underwent fluorescence microscopic analysis [19]. Thereafter, the samples were dried at room temperature and coated with the Vectrashield mounting medium (Sigma-Aldrich, Taufkirchen, Germany). The analysis by epifluorescence microscopy (Axioplan, Zeiss, Oberkochen, Germany) was conducted. The root canal samples with the dentinal tubules were analyzed at 1000-fold, 400-fold, and 100-fold magnifications using a light filter for DAPI (BP 365, FT 395, LP 397 Zeiss, Oberkochen, Germany). The area of the ocular grid allowed the visualization of the total length of the dentinal tubules.

2.6. Scanning Electron Microscopy

Regarding the scanning electron microscopic (SEM) investigation, the other half of the root was used. The sectioned root canal specimens with four roots from each group were transferred into microtubes with 4% glutaraldehyde (Sigma-Aldrich, Taufkirchen, Germany). The fixation with glutaraldehyde was continued for 2.5 h at room temperature. The next step was washing with the PBS for 15 min twice and dehydration with Isopropanol (Carl Roth GmbH Co. KG, Karlsruhe, Germany). Then, chemical drying through iterative transfer into hexamethyldisilazane (HMDS) was performed. The specimens were fixed on SEM stubs and sputtered with gold–palladium. The scanning electron microscopy took place using a Philips ESEM XL 30 in the high-vacuum mode to detect secondary electrons for imaging.

2.7. DATA Evaluation

First, the surface area of each sample was calculated, using the Image J2-Fiji program (Curtis Rueden of UW-Madison LOCI, Madison, WI, USA), in the DAPI images with a 100-fold magnification [22]. Then, the number of bacteria per surface area was calculated visually by two operators using a compact manual cell counter (Fisherbrand™, Schwerte, Germany) in all the DAPI images of each group (apical and medial parts of the root canal). The arithmetic estimation of bacterial penetration depth was performed using the Axio Vision program (Zeiss, Jena, Germany). By that means, the distance between the entrance of the dentinal tubules and the penetrated bacterial cells was measured. Moreover, all cells were counted between the entrance of the dentinal tubules and the deepest penetrated cell in the specimens (Scheme 1).

Collection and preparation of the tooth (*n=120*)

extracted human premolars of mandible without cracks, resorption and opened apex.

decoronation, determination of working length in 15mm, enlarging coronally with Gates Glidden's #5

groove preparation for the splitting of root

↓

Initial chemomechanical preparation

Protaper (F1) and NaCL 0,9% , EDTA 17% (1 min PUI), NaCl 0,9%

↓

Formation of *Enterococcus faecalis* biofilm

sterilization control, preparation (sealing of groove with silicone) for the Eppendorf, **6-week incubation**

↓

Protocol of Disinfection

further mechanical preparation (F4), final irrigation protocols with PUI **6 Groups (*n=20* each)**

↓

Analysis – Visualization of the results

splitting of the tooth, preparation for the **SEM, DAPI-staining**

↓

Data Evaluation

counting of bacteria, statistical analysis

Scheme 1. Flow chart of the experimental setup.

A four-point scoring system was adapted to evaluate the surface profile of the root canal dentine after the application of the disinfection protocols by using the SEM data [23]. The scoring system was defined according to the representative images from the SEM data (Figure 2):

Score 0: Absence of irregularities and dentinal tubules closed.
Score 1: Partially irregular and dentinal tubules partially opened.
Score 2: Damage of the surface and dentinal tubules opened.
Score 3: Severe erosion of the dentinal surface and dentinal tubules widely opened.

2.8. Statistical Analysis

The values were compared by using one-way analysis of variance (ANOVA), followed by a post hoc test (Dunnett'sT3). The Dunnett's T3 test was used to assess the differences between the six groups based on the DAPI data. The level of significance was set at 0.001 for the one-way ANOVA test, with a statistical power of 95%, and at 0.05 for the Dunnett's T3 test, with a statistical power of 80%.

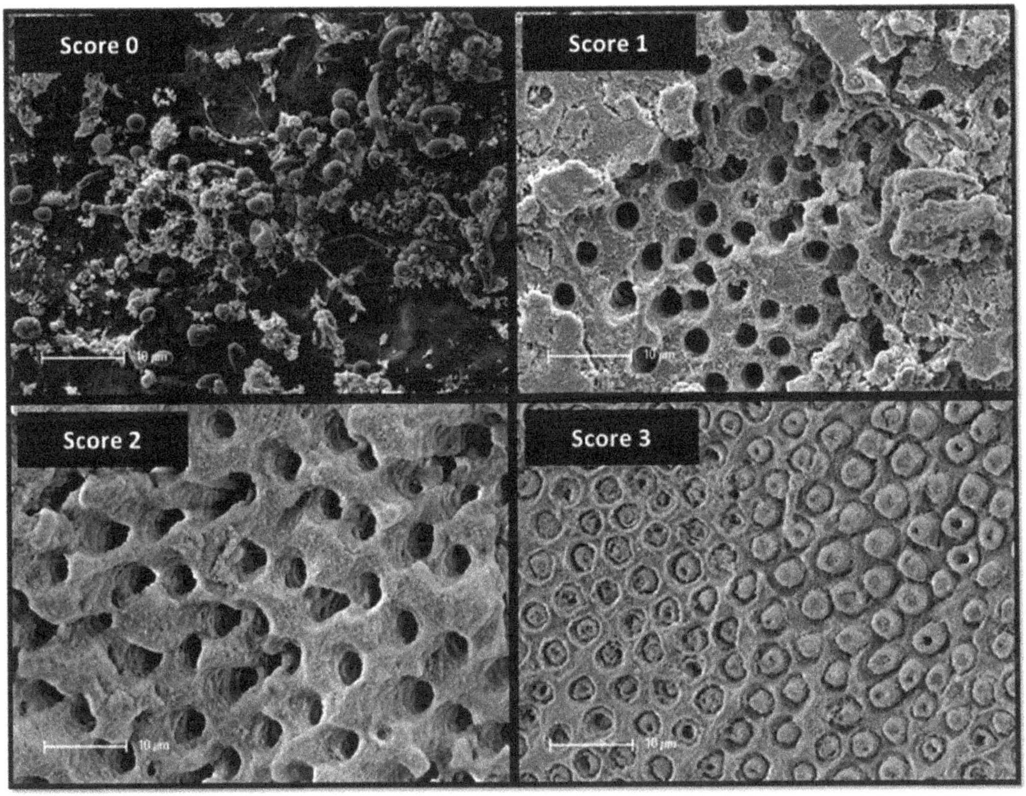

Figure 2. Representative scanning electron micrographs of the surface profiles of root canal walls characterized by a score of 0–3. (5000-fold magnification). Score 0: absence of irregularities of the root canal wall and dentinal tubules closed—sample from the control group; Score 1: partially irregular root canal wall and dentinal tubules partially opened—sample from Group 2; Score 2: damage of the surface and dentinal tubules opened—sample from Group 3; and Score 3: severe erosion of the dentinal surface, dentinal tubules widely opened, and collagen exposed—sample from Group 5.

3. Results

The bacterial colonization in the samples was successful. After six weeks of incubation with *E. faecalis*, the cells had already migrated into the dentinal tubules. The examination with the DAPI method gave insight into the remaining bacterial cells inside the dentinal tubules after the application of the different disinfection protocols. All cells were counted between the entrance of the dentinal tubules and the deepest penetrated cell in the specimens. The overall penetration depths of the deepest remaining *E. faecalis* cells in all groups and both parts of the root canal (medial and apical) were measured to validate the biofilm model. A deeper penetration pattern of *E. faecalis* into the dentinal tubules was observed in the medial part of the root canal compared to the apical part of the root canal in the control group (Figure 3).

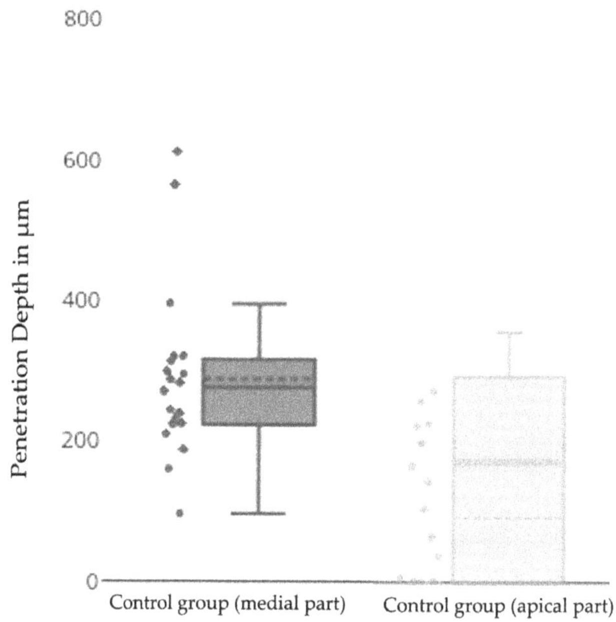

Figure 3. DAPI results: bacterial penetration depth (μm) of the deepest remaining *E. faecalis* cells in the control group from both parts of the root canal (medial and apical) presented in the boxplots. The dots represent all the values, including outliers. Mean scores—mean of value (mv)—are outlined by a dashed line, and median scores are outlined by a straight line: medial part: 289.4 ± 121.554 μm (mv), and apical part: 93.45 ± 118.554 μm (mv).

Bacterial cells were traceable in all specimens by the DAPI method in the control group. The root canals inoculated with *E. faecalis* were heavily infected, and microorganisms were observed in all areas of the dentinal tubules in the control group, even after the instrumentation up to ProTaper file F4 (040/0.06). In general, not only a deeper penetration of *E. faecalis* was observed in the medial part compared to the apical part of the root canal in each group, but the bacterial count was higher as well. More specifically, in the control group, the bacterial count in the medial part was 1492.0 ± 768.4 bact./μm^2, and in the apical part of the root canal, the count was 172.3 ± 222.1 bact./μm^2. The representative images from each group visualize the penetration depth of the deepest remaining bacterial cells, as well as the residual infection of the dentinal tubules with the remaining bacteria in general, after the application of the different disinfection protocols. These images from the DAPI data are representative examples of each group from both parts of the root canal (medial and apical) (Figure 4a,b).

Specifically, less bacteria were detected in Group 2 (NaOCl 3% + EDTA 20% + NaCl 0.9% + CHX 2% with PUI) when compared to Group 4 (CHX 2% + EDTA 20% with PUI) and Group 6 (NaCl 0.9% with PUI). These Groups—4 and 6—yielded comparable amounts of bacteria. Hardly any bacteria were detected in Group 3 (NaOCl 3% + EDTA 20% with PUI) and in Group 5 (NaOCl 3% + MTAD with PUI) after the application of the disinfection protocols. The post hoc comparisons using the Dunnett's T3 test indicated that the mean score for the control group was significantly different from all the other tested groups regarding the medial part of the root canal. However, at the apical part of the root canal, the control group did not significantly differ from Groups 2 (NaOCl 3% + EDTA 20% + NaCl 0.9% + CHX 2% with PUI), 4 (CHX 2% + EDTA 20% with PUI), and 6 (NaCl 0.9% with PUI). There was also no significant difference between Groups 3, 5, and 2 at the medial/apical part of the root canal, but the difference was statistically significant between Groups (3 and

5) and (1, 4, and 6) at the medial part. No statistically significant association could be found between Group 3 and Group 5 (Figure 5).

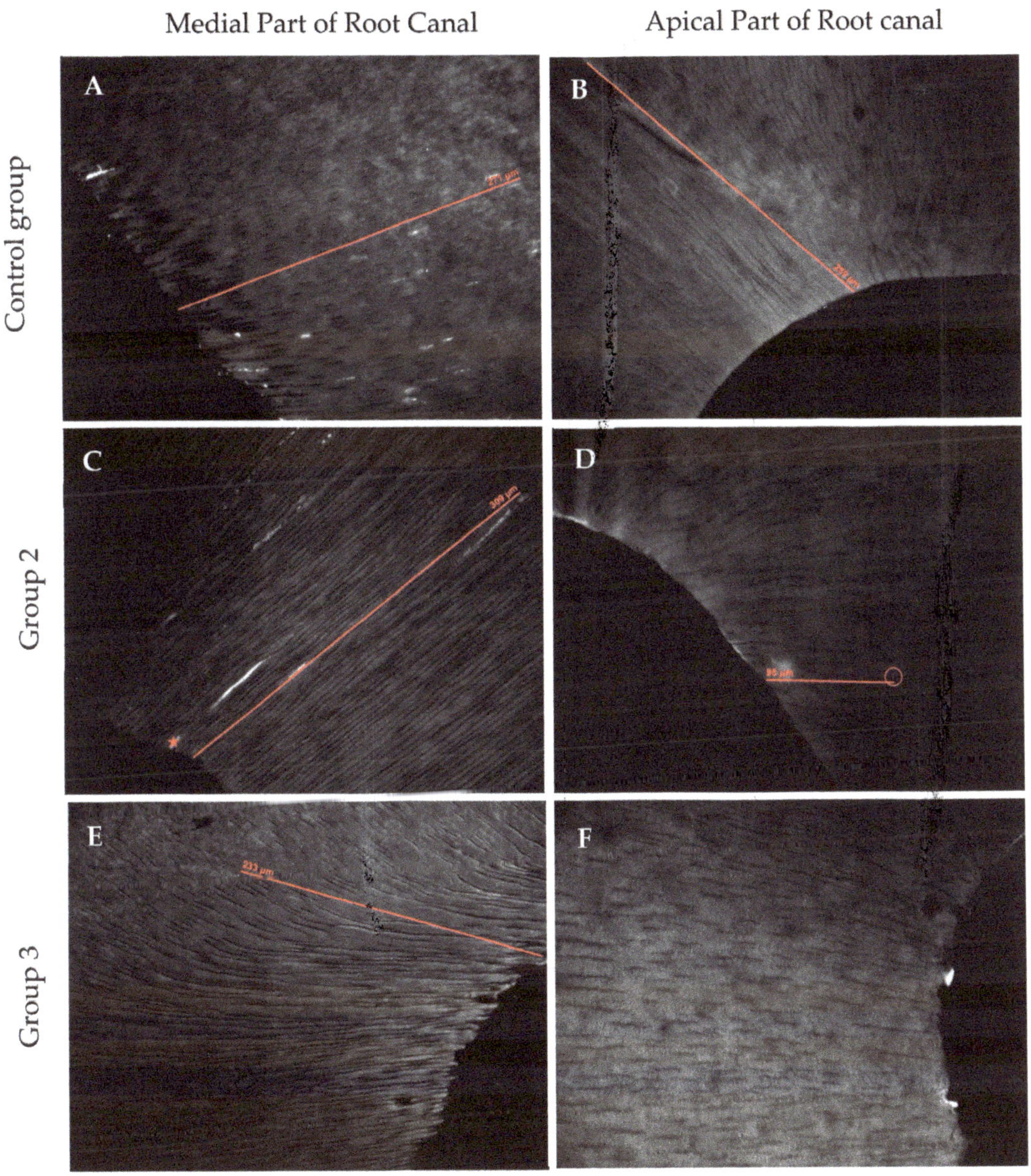

(a)

Figure 4. *Cont.*

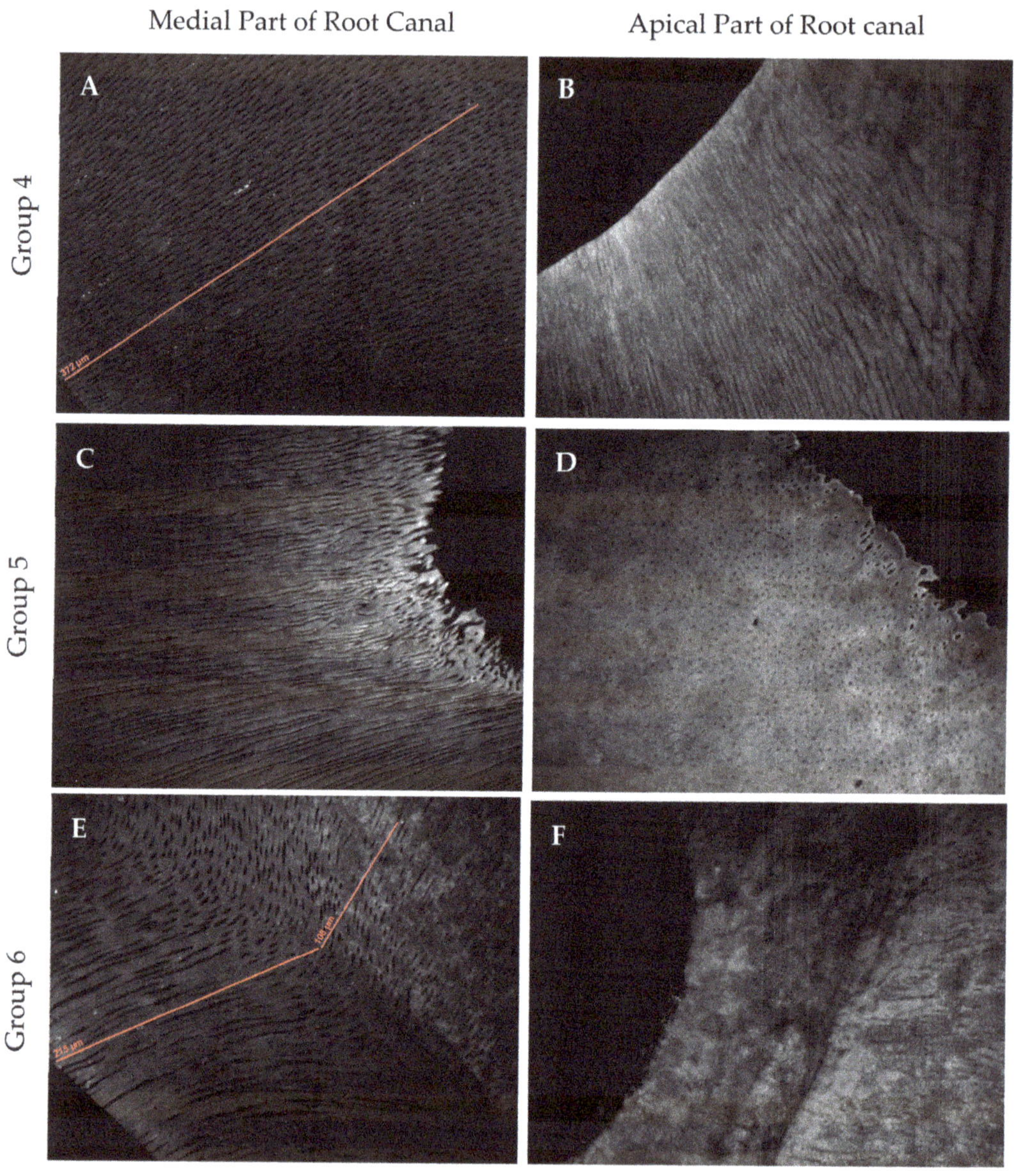

(b)

Figure 4. (**a**) DAPI staining; typical examples for the penetration depth and remaining bacterial colonization/cells for each group in the medial (**A**,**C**,**E**) and apical (**B**,**D**,**F**) parts of the root canal. Measured penetration depth of *E. faecalis* in a representative sample of the control group: in the medial part, it is 271 µm (**A**), and in the apical part, it is 259 µm (**B**). Measured penetration depth in the medial part of Group 2 is 309 µm into the dentinal tubules (**C**) and 95 µm in the apical part of the root canal (**D**). In Group 3, only one specimen sample is found with a penetration depth of 233 µm in the medial part of the root canal (**E**), and there are no bacteria in the apical part of the root canal (**F**). Remaining *E. faecalis* cells (white) are distributed randomly in small aggregates, and the stars (red) outline the artifacts. Group 1: control group; Group 2: NaOCl 3% + EDTA 20% + CHX 2% + NaCl 0.9%

with PUI; and Group 3: NaOCl 3% + EDTA 20% with PUI. (**b**) DAPI staining: typical examples for the penetration depth and remaining bacterial colonization/cells for each group in the medial (**A**,**C**,**E**) and apical (**B**,**D**,**F**) parts of the root canal. No bacteria are detectable in the dentinal tubules of the apical part of Groups 4, 5, and 6 (**B**,**D**,**F**). Measured penetration depth of *E. faecalis* cells in the medial part of Group 6 is 323 μm into the dentinal tubules (E), and it is 372 μm in Group 4 (**A**). In a sample of Group 5 (medial part), the measured penetration depth of the deepest remaining *E. faecalis* cells is 333 μm (**C**). Remaining *E. faecalis* cells (white) are distributed randomly. Group 4: CHX 2% + EDTA 20% with PUI; Group 5: NaOCl 3% + MTAD with PUI; and Group 6: NaCl 0.9% with PUI.

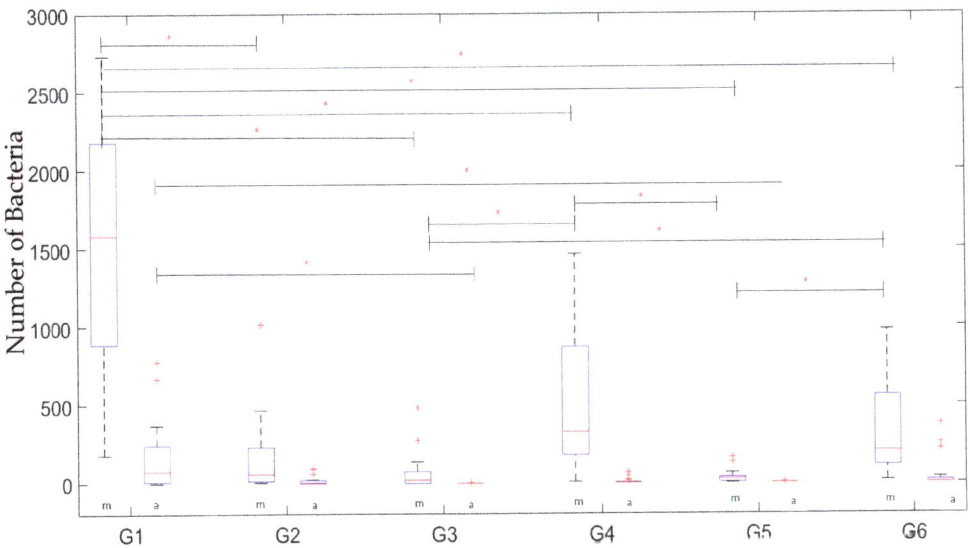

Figure 5. DAPI staining: aggregated presentation of data representing the number of bacteria both at the medial (m) and apical (a) parts of the root canal in all groups. Control Group 1: 1492.0 ± 768.4 bact./μm^2 (m) and 172.3 ± 222.1 bact./μm^2 (a); Group 2 (NaOCl 3% + EDTA 20% + NaCl 0.9% + CHX 2% with PUI): 157.8 ± 239.3 bact./μm^2 (m) and 19.6 ± 31.1 bact./μm^2 (a); Group 3 (NaOCl 3% + EDTA 20% with PUI): 66.2 ± 188.2 bact./μm^2 (m) and 0.5 ± 1.4 bact./μm^2 (a); Group 4 (CHX 2% + EDTA 20% with PUI): 483.8 ± 409.9 bact./μm^2 (m) and 7.9 ± 16.9 bact./μm^2 (a); Group 5 (NaOCl 3% + MTAD with PUI): 33.2 ± 44.3 bact./μm^2 (m) and 0.5 ± 1.6 bact./μm^2 (a); and Group 6 (NaCl 0.9% with PUI): 329.5 ± 275.8 bact./μm^2 (m) and 49.8 ± 109.3 bact./μm^2 (a). Outliers are represented by + (red) in the boxplot. Statistical significances are outlined by * between the various conditions. One-way ANOVA ($p < 0.001$); Dunnett's T3 test ($p < 0.05$).

The SEM data confirmed the results of the DAPI analysis. The visualization of bacteria under the scanning electron microscope (SEM) indicated colonies of *E. faecalis*, especially at the entrances of the dentinal tubules and the root canal surface. It was mainly observed by the SEM how each disinfection protocol had affected the structure of the root dentine.

A considerable alteration of the dentinal structure in the root canal was observed in Groups 3 and 5. This alteration resulted in the erosion of the dentinal ultrastructure. Partial irregularities were observed in Groups 2 and 4. A morphological change of the dentine surface was observed by the SEM in Group 6, as well. The dentinal tubules were partially opened, and there were areas of the dentine surface that were apparently mechanically prepared and instrumented due to the use of the ultrasonic tip (Figure 6).

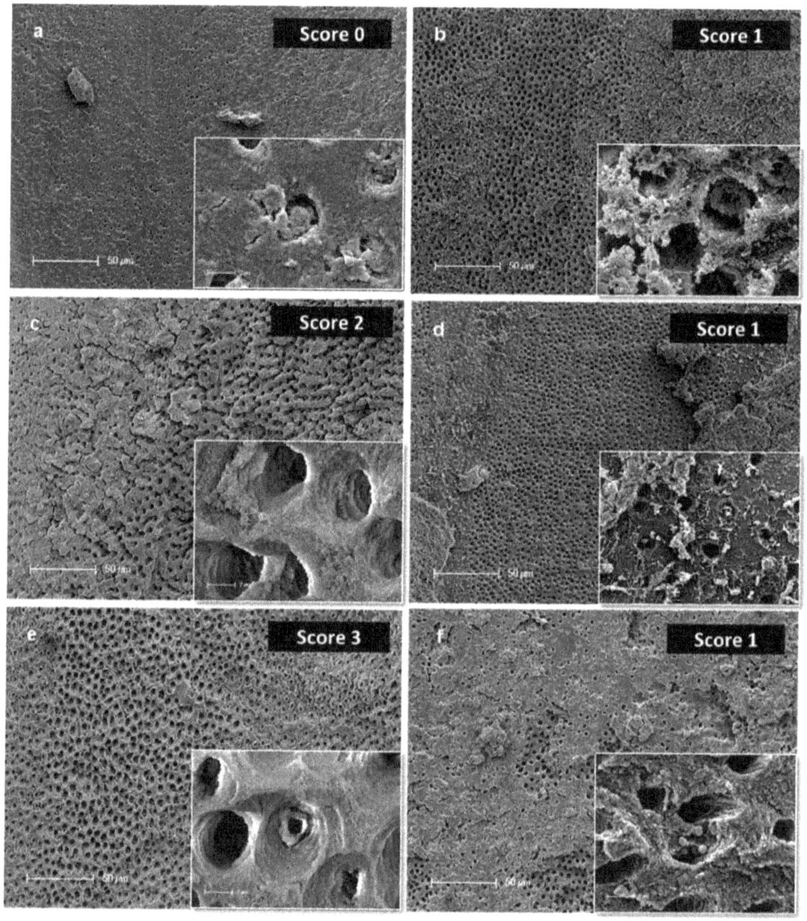

Figure 6. Representative images obtained by SEM from the six groups (1000×, 20,000× magnification) characterized by the scoring system: (**a**) control group: (score 0) no irregularities on the dentine surface; (**b**) Group 2: (Score 1) dentine surface morphology coated with collagen; (**c**) Group 3: (Score 0) demineralization of dentine surface; (**d**) Group 4: (Score 1) limited non-instrumented areas with partially closed dentinal tubules; (**e**) Group 5: (Score 0) severe erosion of the dentinal surface; and (**f**) Group 6: (Score 2) partially irregular dentinal surface.

4. Discussion

With the intention of studying the effect of different disinfection protocols, it was necessary to analyze the cleansing effect of different irrigations and to quantify the remaining bacteria after the decontamination. Therefore, a mono-species biofilm model was established in order to quantify penetrated bacteria into the dentinal tubules of human root dentine after the application of different disinfection protocols. The composition and structure of the endodontic biofilms are highly variable. Although this study was not in vivo and did not mimic complex multispecies endodontic biofilms, the established mono-species *E. faecalis* isolated from a biofilm model with persistent root canal infection provided a well-standardized anatomical and biologically relevant model that allowed the comparison of different disinfection protocols against *E. faecalis* biofilms, as visualized with the DAPI method.

The analytical procedure based on the visualization by DAPI staining was established by the present research group [24,25] to quantify the remaining penetrated *E. faecalis* cells within the dentinal tubules and to evaluate the disinfection effect of five different irrigation protocols. To a certain extent, this technique permits the two-dimensional imaging of bacteria on the root canal surface and inside the dentinal tubules. The examination following DAPI staining gave an additional insight into the remaining penetrated bacterial cells into the dentinal tubules [24,25] and confirmed the penetration of bacteria into the dentinal tubules. Furthermore, the examination with the DAPI method revealed a semi-quantitative assessment of bacteria in the observed areas. Undeniably, the evaluation of the results, specifically the calculation of bacteria, was time consuming. The compact manual cell counter allowed us to register each individual colony prior to testing based on standardized comparisons [26]. Notwithstanding the fact that the method was time-consuming, it provided an accurate estimation of the number of bacterial cells.

A criticism often brought up in relation to all imaging techniques used to evaluate intraradicular biofilms is that the observed areas are to some extent subjectively chosen by the examiner. Using the DAPI method, which also provides a local estimation of remaining bacteria, the cross sections of the root—and not of the whole root canal—were evaluated. Certainly, the field of observation is more extended in contrast to other methods, such as colony-forming units (CFU). Considering that the field of vision under the microscope is limited, most of bacteria remain undetected. Regarding CFU, most of the studies are based on the paper point sampling process for the analysis of bacteria, which is considered controversial since it leads to a biased collection of biofilm material that are readily accessible, while hidden bacteria in dentinal tubules are oftentimes disregarded [27]. A negative culture result by CFU does not necessarily imply a bacteria-free root canal system, as bacteria may be retained in complex areas of the system or into the dentinal tubules embedded within a biofilm, thus being inaccessible to the paper points used for sampling. Therefore, the CFU method in combination with paper points for bacterial identification can result in an underestimation of the bacteria present in an infected root canal [28]. Even though the field of vision in the presented DAPI method in this study is limited, it offers insight into the bacterial contamination within the dentinal tubules and the bacterial penetration depth. Further analysis of the root canal wall using SEM gives additional insight into the bacterial contamination of the root canal surface as well as the alteration of the root dentine caused by endodontic irrigants.

It is apparent that the combination of different microscopic techniques is more likely to facilitate a deeper and more realistic analysis of biofilm architecture in the root canal. The combination with SEM provides more information regarding the visualization of bacteria [29]. SEM also provides information concerning the condition of bacteria and the ultrastructure [30,31]. SEM has been used to visualize the distribution of bacteria on the surface of biofilm in the root canal wall, as well as the penetration inside dentinal tubules [24,32,33]. However, the resulting images are, therefore, only pseudo three-dimensional. As biofilms are multileveled, SEM is unable to assess the full depth of these structures [29,34,35]. Another aspect regarding SEM is that only topographical assessment of the observed structures is possible, which makes it nearly impossible to quantify the bacteria into the root dentine areas and, especially, inside the dentinal tubules. Therefore, only qualitative assessment of observed specimens can be performed with this technique. This is not surprising, considering the field of vision under a scanning electron microscope contains only a few micrograms of dentine. For this reason, previous studies have used SEM in order to visualize bacteria in the root canal and not to quantify them [12,29,36]. However, SEM is a very effective method to analyze ultrastructural surface alterations after an irrigant application. Different studies have already shown the use of SEM to investigate enamel [37,38] and dentine [39,40]. Nevertheless, a direct comparison of irrigant decontamination efficacy using fluorescence microscopy (DAPI) and dentinal surface alteration using SEM was performed for the first time in the present study.

Regarding the viability of the remaining bacteria, the established DAPI method is unable to provide knowledge about the viability of such bacteria. There is no evidence regarding a live/dead staining with the advantages of the DAPI method, which has the ability to detect the viability of penetrated bacteria located within the dentinal tubules of the root canal. The fixation of bacteria—a necessary step in the analysis with the DAPI method—destroys the viability state of cells. In general, it can be assumed that the differentiation of viable and dead bacteria is possible using different live/dead staining methods. These methods represent the viability state during the staining procedure. However, with additional after dye accumulation inside the cells, bacteria, indeed, lose their viability [41].

The model in this study was based on a mono-species biofilm with *E. faecalis* within the root canals and dentinal tubules. *E. faecalis* plays an important role in bacterial biofilm invasion and is considered a suitable model for assessing root canal bacterial penetration. Many in vitro investigations have been undertaken to examine the mechanisms involved in bacterial penetration into dentine and to visualize this infection in dentinal tubules [12,42]. Confocal laser scanning microscopy (CLSM) and the DAPI method—which was used in this study—allow the most precise assessment of bacterial penetration in the dentinal tubules and generate less risk of creating artifacts, when compared to SEM [43]. Most of studies with CLSM use colony-forming units (CFU) in order to quantify bacteria that have invaded into dentinal tubules. Considering the drawbacks of CFU, the DAPI method can be considered as an alternative visualization method for quantifying bacterial penetration.

In this study, the effectiveness of the five different disinfection protocols was examined in comparison to the control group, and it was concluded that the disinfection protocols in Groups 3 (NaOCl 3% + EDTA 20% with PUI) and 5 (NaOCl 3% + MTAD with PUI) were the most effective ($p < 0.001$). There was a significant difference in the bacterial count between both Groups 3 and 5 and all the other groups in the medial part of the root canal ($p < 0.05$). Although PUI improved the effectiveness of conventional irrigation, no significant difference was detected between the control group (only instrumentation and conventional irrigation with NaCl 0.9%) and Group 6 (instrumentation and passive ultrasonic irrigation with NaCl 0.9%) based on the Dunnett's T3 test ($p < 0.05$). Obviously, the irrigants play an important role in the decontamination of *E. faecalis* biofilm, but the use of passive ultrasonic irrigation also enhances bacterial reduction from the root canal systems when compared to other methods of irrigant activation and conventional syringe irrigation [6,44–46].

Both groups (Groups 3 and 5) revealed the best results concerning decontamination and biofilm dissolution capacity. NaOCl is appropriate as an irrigant because it is effective in disrupting biofilm [47]. However, structural deformations/alterations were observed in the dentine ultrastructure. The use of NaOCl in combination with additional irrigants as the final irrigation provokes severe structural changes in the dentinal collagen. These phenomena has also been observed in previous ultramorphological studies [48,49]. The erosion of the dentine by proteolytic degradation is followed by the formation of fragile, spongy-like root dentine. The worst-case scenario is root fracture due to the weakening of the root. Therefore, it is a fine line between removing too much dental tissue, which would strongly weaken the root, and leaving the infected dental tissue in the root canal, which would reduce the possibility of achieving the best decontamination effects.

Intensive research is being conducted to develop disinfection protocols for the root canal system. The ideal protocol of disinfection does not exist yet. In the near future, nanotechnology might be applied to the endodontic field. In endodontics, there are no techniques that promote total anti-biofilm removal while simultaneously do not affect the root dentine ultrastructure. Nanomaterials and nanocarriers could open new opportunities for the removal of biofilms and repair of tooth structure [50].

5. Conclusions

In conclusion, a clinically relevant *E. faecalis* biofilm model for in vitro studies—based on visualization by DAPI staining—was established. The results of the present study suggest that the combination of 3% NaOCl under passive ultrasonic irrigation with an additional final irrigant, such as 20% EDTA or MTAD, is the most effective disinfection protocol against *E. faecalis* biofilm. Yet, the dentinal root canal surface is altered the most after the application of 3% NaOCl with 20% EDTA or MTAD. The combination of epifluorescence microscopy with DAPI staining and scanning electron microscopy (SEM) is a novel approach of the present research group to visualize and quantify the decontamination effects of different endodontic irrigants by evaluating the remaining bacteria on the root canal surface and within the dentinal tubules. At the same time, the penetration depth of bacteria and the bacterial ultrastructure/condition, as well as the alteration of the dentine due to the endodontic irrigants, can be evaluated.

Author Contributions: Conceptualization, C.H., M.-T.K. and M.D. (Martin Dannemann); data curation, M.D. (Maria Dede); formal analysis, J.N.; investigation, M.D. (Maria Dede); methodology, M.D. (Maria Dede), S.B., C.H. and M.-T.K.; project administration, M.D. (Martin Dannemann) and M.-T.K.; supervision, C.H. and M.-T.K.; validation, M.D. (Maria Dede) and S.B.; visualization, M.D. (Maria Dede); writing—original draft, M.D. (Maria Dede); writing—review and editing, M.D. (Maria Dede), C.H. and M.-T.K. All authors have read and agreed to the published version of the manuscript.

Funding: This research was funded by the DFG (German Research Foundation), grant number (WE 5838/1-1 and WE 5838/1-2, DA 1701/1-1 and DA 1701/1-2).

Institutional Review Board Statement: Approval by the ethics committee of the University of Freiburg (140/09) for the isolation of *E. faecalis*.

Data Availability Statement: The data presented in this study are available from the corresponding author upon request.

Acknowledgments: The Article Processing Charges (APC) were funded by the joint publication funds of the TU Dresden, including Carl Gustav Carus Faculty of Medicine, and the SLUB Dresden as well as the Open Access Publication Funding of the DFG.

Conflicts of Interest: The authors declare no conflict of interest. Except as disclosed above, the authors certify that they have no commercial associations that might represent a conflict of interest in connection with the submitted manuscript. The funders had no role in the design of the study; in the collection, analyses, or interpretation of data; in the writing of the manuscript; or in the decision to publish the results.

References

1. Ramachandran Nair, P.N. Light and Electron Microscopic Studies of Root Canal Flora and Periapical Lesions. *J. Endod.* **1987**, *13*, 29–39. [CrossRef] [PubMed]
2. Ordinola-Zapata, R.; Noblett, W.C.; Perez-Ron, A.; Ye, Z.; Vera, J. Present Status and Future Directions of Intracanal Medicaments. *Int. Endod. J.* **2022**, *55*, 613–636. [CrossRef] [PubMed]
3. Plotino, G.; Cortese, T.; Grande, N.M.; Leonardi, D.P.; Di Giorgio, G.; Testarelli, L.; Gambarini, G. New Technologies to Improve Root Canal Disinfection. *Braz. Dent. J.* **2016**, *27*, 3–8. [CrossRef] [PubMed]
4. Cunningham, W.; Martin, H. A Scanning Electron Microscope Evaluation of Root Canal Débridement with the Endosonic Ultrasonic Synergistic System. *Oral Surg. Oral Med. Oral Pathol.* **1982**, *53*, 527–531. [CrossRef] [PubMed]
5. Lee, S.; Wu, M.; Wesselink, P. The Efficacy of Ultrasonic Irrigation to Remove Artificially Placed Dentine Debris from Differ-ent-sized Simulated Plastic Root Canals. *Int. Endod. J.* **2004**, *37*, 607–612. [CrossRef]
6. Van der Sluis, L.W.M.; Versluis, M.; Wu, M.K.; Wesselink, P.R. Passive Ultrasonic Irrigation of the Root Canal: A Review of the Literature. *Int. Endod. J.* **2007**, *40*, 415–426. [CrossRef]
7. Love, R.M. *Enterococcus faecalis*—A Mechanism for Its Role in Endodontic Failure. *Int. Endod. J.* **2001**, *34*, 399–405. [CrossRef]
8. Barbosa-Ribeiro, M.; De-Jesus-Soares, A.; Zaia, A.A.; Ferraz, C.C.; Almeida, J.F.; Gomes, B.P. Antimicrobial Susceptibility and Characterization of Virulence Genes of *Enterococcus faecalis* Isolates from Teeth with Failure of the Endodontic Treatment. *J. Endod.* **2016**, *42*, 1022–1028. [CrossRef] [PubMed]
9. Siqueira, J.; Rôças, I.; Rosado, A. Investigation of Bacterial Communities Associated with Asymptomatic and Symptomatic Endodontic Infections by Denaturing Gradient Gel Electrophoresis Fingerprinting Approach. *Oral Microbiol. Immunol.* **2004**, *19*, 363–370. [CrossRef]
10. Andrewes, F.; Horder, T. A Study of the streptococci pathogenic for man. *Lancet* **1906**, *168*, 852–855. [CrossRef]
11. Vidana, R.; Billström, H.; Sullivan, Å.; Ahlquist, M.; Lund, B. *Enterococcus faecalis* Infection in Root Canals—Host-derived or Exogenous Source? *Lett. Appl. Microbiol.* **2011**, *52*, 109–115. [CrossRef]

12. Ørstavik, D.; Haapasalo, M. Disinfection by Endodontic Irrigants and Dressings of Experimentally Infected Dentinal Tubules. *Endod. Dent. Traumatol.* **1990**, *6*, 142–149. [CrossRef] [PubMed]
13. Siqueirajr, J.; Rôças, I.N.; Favieri, A.; Lima, K.C. Chemomechanical Reduction of the Bacterial Population in the Root Canal after Instrumentation and Irrigation with 1%, 2.5%, and 5.25% Sodium Hypochlorite. *J. Endod.* **2000**, *26*, 331–334. [CrossRef] [PubMed]
14. Pladisai, P.; Ampornaramveth, R.S.; Chivatxaranukul, P. Effectiveness of Different Disinfection Protocols on the Reduction of Bacteria in *Enterococcus faecalis* Biofilm in Teeth with Large Root Canals. *J. Endod.* **2016**, *42*, 460–464. [CrossRef]
15. Gründling, G.L.; Zechin, J.G.; Jardim, W.M.; Oliveira, S.D.; De Figueiredo, J.P. Effect of Ultrasonics on *Enterococcus faecalis* Biofilm in a Bovine Tooth Model. *J. Endod.* **2011**, *37*, 1128–1133. [CrossRef] [PubMed]
16. Estrela, C.; Sydney, G.; de Figueiredo, J.P. A Model System to Study Antimicrobial Strategies in Endodontic Biofilms. *J. Appl. Oral Sci.* **2009**, *17*, 87–91. [CrossRef] [PubMed]
17. Teves, A.; Blanco, D.; Casaretto, M.; Torres, J.; Alvarado, D.E.; Coaguila-Llerena, H.; Faria, G.; Jaramillo, D.E. Multispecies Biofilm Removal by XP-endo Finisher and Passive Ultrasonic Irrigation: A scanning Electron Microscopy Study. *Aust. Endod. J.* **2022**, *48*, 91–97. [CrossRef] [PubMed]
18. Morikawa, K.; Yanagida, M. Visualization of Individual DNA Molecules in Solution by Light Microscopy: DAPI Staining Method. *J. Biochem.* **1981**, *89*, 693–696. [CrossRef] [PubMed]
19. Kensche, A.; Basche, S.; Bowen, W.H.; Hannig, M.; Hannig, C. Fluorescence Microscopic Visualization of Non Cellular Components during Initial Bioadhesion in situ. *Arch. Oral Biol.* **2013**, *58*, 1271–1281. [CrossRef]
20. Hannig, C.; Hannig, M.; Rehmer, O.; Braun, G.; Hellwig, E.; Al-Ahmad, A. Fluorescence Microscopic Visualization and Quantification of Initial Bacterial Colonization on Enamel in situ. *Arch. Oral Biol.* **2007**, *52*, 1048–1056. [CrossRef]
21. Jung, D.J.; Al-Ahmad, A.; Follo, M.; Spitzmüller, B.; Hoth-Hannig, W.; Hannig, M.; Hannig, C. Visualization of Initial Bacterial Colonization on Dentine and Enamel in situ. *J. Microbiol. Methods* **2010**, *81*, 166–174. [CrossRef] [PubMed]
22. Schindelin, J.; Arganda-Carreras, I.; Frise, E.; Kaynig, V.; Longair, M.; Pietzsch, T.; Preibisch, S.; Rueden, C.; Saalfeld, S.; Schmid, B.; et al. Fiji: An Open-source Platform for Biological-image Analysis. *Nat. Methods* **2012**, *9*, 676–682. [CrossRef]
23. Prati, C.; Foschi, F.; Nucci, C.; Montebugnoli, L.; Marchionni, S. Appearance of the Root Canal Walls after Preparation with NiTi Rotary Instruments: A comparative SEM Investigation. *Clin. Oral Investig.* **2004**, *8*, 102–110. [CrossRef] [PubMed]
24. Kirsch, J.; Basche, S.; Neunzehn, J.; Dede, M.; Dannemann, M.; Hannig, C.; Weber, M.-T. Is it Really Penetration? Locomotion of Devitalized *Enterococcus faecalis* Cells within Dentinal Tubules of Bovine Teeth. *Arch. Oral Biol.* **2017**, *83*, 289–296. [CrossRef] [PubMed]
25. Kirsch, J.; Basche, S.; Neunzehn, J.; Dede, M.; Dannemann, M.; Hannig, C.; Weber, M.-T. Is it Really Penetration? Part 2. Locomotion of *Enterococcus faecalis* Cells within Dentinal Tubules of Bovine Teeth. *Clin. Oral Investig.* **2019**, *23*, 4325–4334. [CrossRef] [PubMed]
26. Kubista, M.; Aakerman, B.; Norden, B. Characterization of Interaction between DNA and 4′,6-diamidino-2-phenylindole by Optical Spectroscopy. *Biochemistry* **1987**, *26*, 4545–4553. [CrossRef] [PubMed]
27. Sterzenbach, T.; Pioch, A.; Dannemann, M.; Hannig, C.; Weber, M.-T. Quantification of Bacterial Colonization in Dental Hard Tissues Using Optimized Molecular Biological Methods. *Front. Genet.* **2020**, *11*, 599137. [CrossRef] [PubMed]
28. Siqueira, J.F.; Rôças, I.N. PCR Methodology As a Valuable Tool for Identification of Endodontic Pathogens. *J. Dent.* **2003**, *31*, 333–339. [CrossRef]
29. Schaudinn, C.; Carr, G.; Gorur, A.; Jaramillo, D.; Costerton, J.W.; Webster, P. Imaging of Endodontic Biofilms by Combined Microscopy (FISH/cLSM—SEM). *J. Microsc.* **2009**, *235*, 124–127. [CrossRef]
30. Huque, J.; Kota, K.; Yamaga, M.; Iwaku, M.; Hoshino, E. Bacterial Eradication from Root Dentine by Ultrasonic Irrigation with Sodium Hypochlorite. *Int. Endod. J.* **1998**, *31*, 242–250. [CrossRef]
31. Fritz, B.; Stavnsbjerg, C.; Markvart, M.; Damgaard, P.D.B.; Nielsen, S.H.; Bjørndal, L.; Qvortrup, K.; Bjarnsholt, T. Shotgun Sequencing of Clinical Biofilm following Scanning Electron Microscopy Identifies Bacterial Community Composition. *Pathog. Dis.* **2019**, *77*, ftz013. [CrossRef]
32. Al-Nazhan, S.; Al-Sulaiman, A.; Al-Rasheed, F.; Alnajjar, F.; Al-Abdulwahab, B.; Al-Badah, A. Microorganism Penetration in Dentinal Tubules of Instrumented and Retreated Root Canal Walls. In vitro SEM Study. *Restor. Dent. Endod.* **2014**, *39*, 258–264. [CrossRef]
33. Yoo, Y.-J.; Kim, A.R.; Perinpanayagam, H.; Han, S.H.; Kum, K.-Y. *Candida albicans* Virulence Factors and Pathogenicity for Endodontic Infections. *Microorganisms* **2020**, *8*, 1300. [CrossRef]
34. Kishen, A.; George, S.; Kumar, R. *Enterococcus faecalis*-Mediated Biomineralized Biofilm Formation on Root Canal Dentine in vitro. *J. Biomed. Mater. Res. A* **2006**, *77*, 406–415. [CrossRef] [PubMed]
35. Seet, A.N.; Zilm, P.S.; Gully, N.J.; Cathro, P.R. Qualitative Comparison of Sonic or Laser Energisation of 4% Sodium Hypochlorite on an *Enterococcus faecalis* Biofilm Grown in vitro. *Aust. Endod. J.* **2012**, *38*, 100–106. [CrossRef]
36. Wanicharat, W.; Wanachantararak, P.; Poomanee, W.; Leelapornpisid, P.; Leelapornpisid, W. Potential of Bouea Macrophylla kernel Extract as an Intracanal Medicament against Mixed-species Bacterial-fungal Biofilm. An In Vitro and Ex Vivo Study. *Arch. Oral Biol.* **2022**, *143*, 105539. [CrossRef]
37. Coceska, E.; Gjorgievska, E.; Coleman, N.J.; Gabric, D.; Slipper, I.J.; Stevanovic, M.; Nicholson, J.W. Enamel Alteration following Tooth Bleaching and Remineralization. *J. Microsc.* **2016**, *262*, 232–244. [CrossRef]

38. Suriyasangpetch, S.; Sivavong, P.; Niyatiwatchanchai, B.; Osathanon, T.; Gorwong, P.; Pianmee, C.; Nantanapiboon, D. Effect of Whitening Toothpaste on Surface Roughness and Colour Alteration of Artificially Extrinsic Stained Human Enamel: In Vitro Study. *Dent. J.* **2022**, *10*, 191. [CrossRef] [PubMed]
39. Montes, M.A.J.R.; De Goes, M.F.; Sinhoreti, M. The In Vitro Morphological Effects of Some Current Pre-treatments on Dentin Surface: A SEM Evaluation. *Oper. Dent.* **2005**, *30*, 201–212. [PubMed]
40. Santosh, A.; Kakade, A.; Mali, S.; Lalwani, R.; Badnaware, S.; Deshmukh, B.; Shetty, H. Scanning Electron Microscopic Observation of Primary Tooth Dentin Following Final Irrigation with 95% Ethanol: An In Vitro Study. *J. Indian Soc. Pedod. Prev. Dent.* **2019**, *37*, 271–274. [CrossRef]
41. Tawakoli, P.N.; Al-Ahmad, A.; Hoth-Hannig, W.; Hannig, M.; Hannig, C. Comparison of Different Live/Dead Stainings for Detection and Quantification of Adherent Microorganisms in the Initial Oral Biofilm. *Clin. Oral Investig.* **2013**, *17*, 841–850. [CrossRef]
42. Love, R.M.; McMillan, M.D.; Park, Y.; Jenkinson, H.F. Coinvasion of Dentinal Tubules by *Porphyromonas gingivalis* and *Streptococcus gordonii* Depends upon Binding Specificity of Streptococcal Antigen I/II Adhesin. *Infect. Immun.* **2000**, *68*, 1359–1365. [CrossRef] [PubMed]
43. Tsesis, I.; Elbahary, S.; Venezia, N.B.; Rosen, E. Bacterial Colonization in the Apical Part of Extracted Human Teeth following Root-end Resection and Filling: A Confocal Laser Scanning Microscopy Study. *Clin. Oral Investig.* **2018**, *22*, 267–274. [CrossRef]
44. Mozo, S.; Llena, C.; Forner, L. Review of Ultrasonic Irrigation in Endodontics: Increasing Action of Irrigating Solutions. *Med. Oral Patol. Oral Cir. Bucal.* **2012**, *17*, e512–e516. [CrossRef]
45. Nagendrababu, V.; Jayaraman, J.; Suresh, A.; Kalyanasundaram, S.; Neelakantan, P. Effectiveness of Ultrasonically Activated Irrigation on Root Canal Disinfection: A Systematic Review of In Vitro Studies. *Clin. Oral Investig.* **2018**, *22*, 655–670. [CrossRef] [PubMed]
46. Nair, R.M.; Jayasree, S.; Poornima, E.S.; Ashique, M. Comparative Evaluation of Antimicrobial Efficacy on *Enterococcus faecalis* and Smear Layer Removal in Curved Canals by Different Irrigation Techniques: An In Vitro Study. *J. Conserv. Dent.* **2022**, *25*, 409. [CrossRef]
47. Rodrigues, C.T.; de Andrade, F.B.; De Vasconcelos, L.R.S.M.; Midena, R.Z.; Pereira, T.C.; Kuga, M.C.; Duarte, M.A.H.; Bernardineli, N. Antibacterial Properties of Silver Nanoparticles As a Root Canal Irrigant against *Enterococcus faecalis* Biofilm and Infected Dentinal Tubules. *Int. Endod. J.* **2018**, *51*, 901–911. [CrossRef] [PubMed]
48. Moreira, D.M.; Almeida, J.F.; Ferraz, C.C.; Gomes, B.P.; Line, S.R.; Zaia, A.A. Structural Analysis of Bovine Root Dentin after Use of Different Endodontics Auxiliary Chemical Substances. *J. Endod.* **2009**, *35*, 1023–1027. [CrossRef]
49. Wagner, M.H.; da Rosa, R.A.; de Figueiredo, J.A.P.; Duarte, M.A.H.; Pereira, J.R.; Só, M.V.R. Final Irrigation Protocols May Affect Intraradicular Dentin Ultrastructure. *Clin. Oral Investig.* **2016**, *21*, 2173–2182. [CrossRef]
50. Diogo, P.; Amparo, F.; Faustino, M.; Palma, P.J.; Rai, A.; Graça, P.M.S.; Neves, M.; Miguel Santos, J. May Carriers at Nanoscale Improve the Endodontic's Future? *Adv. Drug Deliv. Rev.* **2023**, *195*, 114731. [CrossRef]

Journal of *Functional Biomaterials*

MDPI

Article

Effect of Intracoronal Sealing Biomaterials on the Histological Outcome of Endodontic Revitalisation in Immature Sheep Teeth—A Pilot Study

Elanagai Rathinam [1,*], Sivaprakash Rajasekharan [1], Heidi Declercq [2,3], Christian Vanhove [4], Peter De Coster [5] and Luc Martens [1]

1 ELOHA (Equal Lifelong Oral Health for All) Research Group, Paediatric Dentistry, Oral Health Sciences, Ghent University Hospital, 9000 Ghent, Belgium
2 Tissue Engineering and Biomaterials Group, Department of Human Structure and Repair, Ghent University Hospital, Ghent University, 9000 Ghent, Belgium
3 Tissue Engineering Laboratory, Department of Development and Regeneration, KU Leuven, 8500 Kortrijk, Belgium
4 Medical Imaging & Signal Processing, Infinity Laboratory, Ghent University Hospital, Ghent University, 9000 Ghent, Belgium
5 Department of Reconstructive Dentistry and Oral Biology, Dental School, Ghent University Hospital, Ghent University, 9000 Ghent, Belgium
* Correspondence: elanagai.rathinam@ugent.be

Abstract: The influence of intracoronal sealing biomaterials on the newly formed regenerative tissue after endodontic revitalisation therapy remains unexplored. The objective of this study was to compare the gene expression profiles of two different tricalcium silicate-based biomaterials alongside the histological outcomes of endodontic revitalisation therapy in immature sheep teeth. The messenger RNA expression of TGF-β, BMP2, BGLAP, VEGFA, WNT5A, MMP1, TNF-α and SMAD6 was evaluated after 1 day with qRT-PCR. For evaluation of histological outcomes, revitalisation therapy was performed using Biodentine (n = 4) or ProRoot white mineral trioxide aggregate (WMTA) (n = 4) in immature sheep according to the European Society of Endodontology position statement. After 6 months' follow-up, one tooth in the Biodentine group was lost to avulsion. Histologically, extent of inflammation, presence or absence of tissue with cellularity and vascularity inside the pulp space, area of tissue with cellularity and vascularity, length of odontoblast lining attached to the dentinal wall, number and area of blood vessels and area of empty root canal space were measured by two independent investigators. All continuous data were subjected to statistical analysis using Wilcoxon matched-pairs signed rank test at a significance level of $p < 0.05$. Biodentine and ProRoot WMTA upregulated the genes responsible for odontoblast differentiation, mineralisation and angiogenesis. Biodentine induced the formation of a significantly larger area of neoformed tissue with cellularity, vascularity and increased length of odontoblast lining attached to the dentinal walls compared to ProRoot WMTA ($p < 0.05$), but future studies with larger sample size and adequate power as estimated by the results of this pilot study would confirm the effect of intracoronal sealing biomaterials on the histological outcome of endodontic revitalisation.

Keywords: Biodentine; dental pulp; immature teeth; mineral trioxide aggregate; MTA; regenerative endodontics; revitalisation

1. Introduction

Revitalisation therapy is based on the theory that undifferentiated stem cells and progenitor cells in the apical papilla of immature permanent teeth can be recruited in the root canal to regenerate functional dentine–pulp complex, induce deposition of mineralized tissue on the root canal wall and replace the necrotic pulp tissue lost due to caries, trauma or developmental anomalies [1,2]. Revitalisation therapy hypothesizes that in contrast to apexification therapy, apexogenesis will occur alongside the strengthening of the root canal walls and consequently prevent cervical fractures and increase the survival rate [3]. An important goal of revitalisation therapy is to form new tissue with the same histological characteristics to recapitulate the lost dental pulp. A majority of the animal and human studies, however, report the formation of pulp-like, cementum-like, bone-like, periodontal ligament-like or fibrous connective tissue [4].

The meticulous combination and interplay of three key elements, namely, stem cells, scaffold and bioactive molecules is referred to as the triad of tissue engineering [5,6]. The intentional manipulation of these three components of the tissue engineering triad has been investigated, but thus far, no protocol can achieve predictable endodontic tissue regeneration [7]. Stem/progenitor cells recruited from the apical region of the tooth are derived from peripheral blood, periodontal ligament, bone marrow or granulation tissue [8]. This has been proposed as a reason for the varied histology of the newly formed tissue [9]. Induced blood clot is the traditionally used scaffold, but certain studies suggest that blood clot might not be a stable scaffold to support tissue regeneration at its initial phase [10]. Alternative natural or synthetic scaffolds such as polylactic acid, polyglycolic acid, polycaprolactone, collagen, fibrin, chitosan, hyaluronic acid, poly(lactic-co-glycolic acid), peptide hydrogels, platelet-rich plasma and platelet-rich fibrin have been experimented with as possible scaffolds to enhance the histological outcome of revitalisation therapy [4,11].

Growth factors are critical signalling molecules that guide the stem cells towards differentiation [12]. These bioactive molecules are delivered at the site of regeneration from various sources such as demineralized dentine matrix fossilized with growth factors [13], paracrine/trophic effects of the stem cells [14,15], scaffolds impregnated with growth factors [16] and the molecular signalling induced with the biomaterial used as the intracoronal barrier. Tricalcium silicate cements (TSCs) are the most preferred biomaterials used for coronal seal during revitalisation therapy [17]. TSCs have been shown to induce the dental pulp stem cell differentiation regulated by a complex network of signalling molecules, pathways, receptors and transcription control systems [18,19].

Extensive literature is available on the variables that affect the histological outcome of revitalisation therapy such as age of the patient [20,21], diameter of open apex [21], extent of pulp necrosis [22,23], residual pulp in the canal [24], residual bacteria [25], disinfectants [26,27], irrigants [9,28,29], intracanal medicaments [28], stem cells [30] and scaffolds [4]. However, there are no studies comparing the influence of the intracoronal biomaterial on the histology of the newly formed regenerative tissue. Hence, the objective of this study was to compare the gene expression profiles of two different tricalcium silicate based biomaterials, namely, Biodentine and ProRoot white mineral trioxide aggregate (WMTA) alongside histological outcomes of revitalisation therapy in immature sheep teeth. The null hypothesis was that there is no difference in the histologic outcome between revitalisation therapy using Biodentine or ProRoot WMTA.

2. Materials and Methods

2.1. Cell Isolation

Human dental pulp stem cells (hDPSCs) were isolated from extracted unerupted human third molars with enzyme digestion [31]. Informed consent was collected from all patients, and ethical approval was obtained from the Ethical Committee of University hospital, Ghent, Belgium, according to the laws of the ICH Good Clinical Practice (GE11-LM-go-2006/57). The teeth were cleaned, cut with a bone cutter at the cemento–enamel junction to remove the pulp tissue and digested with type 1 collagenase and dispase. Small pieces of dental pulp tissue were transferred into enzyme solution for 1 h at 37 °C and vortexed every 30 min to break up the tissue. Thereafter, large cell aggregates were removed, and single-cell suspension was obtained via filtering through a cell-strainer (70 μM). The single-cell suspension was centrifuged at 1200 rpm for 5 min at room temperature. Supernatants were pipetted, and the pellet was resuspended in 1 mL basic medium (fetal bovine serum) to terminate enzymatic dissocation. Cell suspension was cultured in a 25 cm^2 flask in Alpha-modified Eagles medium (α-MEM, Sigma-Aldrich, Overijse, Belgium) with 10% fetal bovine serum and antibiotics (100 U/mL penicillin and 100 mg/mL streptomycin) at 37 °C and 5% CO_2. Then, culture medium was changed every 3 days until the cell confluency was achieved. Flow cytometry analysis was performed to identify the purity of the stem cell culture obtained. Purity of the custom prepared stem cells was determined as 96% using the mesenchymal stem cell markers CD90 and CD105.

2.2. Sample Preparation

ProRoot white MTA (WMTA) (Dentsply, Tulsa; Tulsa, OK, USA) and Biodentine (Septodont, Saint Maur des Fosses, France) were mixed according to the manufacturer's instructions and allowed to set in Teflon moulds of 2 mm height and 8 mm diameter (n = 3 per group). The samples were allowed to set in 100% relative humidity at 37 °C for 3 h. The samples were removed from the mould and UV-sterilized for 30 min on each side. ProRoot white MTA and Biodentine were then placed in Transwell inserts of 0.4 μm in a 6-well plate seeded with hDPSCs at a density of 5×10^5 cells/well and maintained at 37 °C and 5% CO_2 for 1 day.

2.3. qRT-PCR Analysis

qRT-PCR analyses were completed after 1 day. cDNA was produced and amplified using a reverse transcriptome kit (QuantiTect Reverse Transcription kit, Qiagen, Hilden, Germany). Target cDNA was amplified using specific primer pairs. qRT-PCR was performed using the Sensimix SYBR No-ROX Kit (Bioline, London, UK) on a Light cycler 480 System (Roche Life Science, Penzberg, Germany). Samples were normalized using qBasePlus (Biogazelle NV, Zwijnaarde, Belgium) against at least three of the following genes: *Rpl13a, Eif4b, B2m, Actb*, or *Gapdh* as described previously [32]. Details of the specific primers used for gene expression analysis are provided in Table 1.

Table 1. Details of the primers used in this study.

Target Gene	Primer	
	Forward	Reverse
GAPDH	CTACCAGTGCAAAGAGCCCA	TGGTCATCAACCCTTCCACG
ACTB	CTTCGCGGGCGACGAT	CCACATAGGAATCCTTCTGACC
B2M	ACTTAGAGGTGGGGAGCAGA	GCCCTTTACACTGTGAGCC
EIF4b	GTGCGTTTACCACGTGAACC	CGTGCATCCTGGTCTGACTT
RPH3a	CTGGTCCGAGTTTTCTCCGC	TTCTTTATCATTTGATTGAAGGGGC
TGF-β1	AGGGCTACCATGCCAACTTC	GACACAGAGATCCGCAGTCC
BMP2	AGTCCTGATGAGCATGAGCC	CTCACCTATCTGTATACTGC
BGLAP	CTCACACTCCTCGCCCTAT	TCTCTTCACTACCTCGCTGC
VEGFA	ATGCGGATCAAACCTCACCA	CACCAACGTACACGCTCCAG
WNT5a	AAGCAGACGTTTCGGCTACA	TTTCCAACGTCCATCAGCGA
MMP1	CCCAGCGACTCTAGAAACACA	CTGCTTGACCCTCAGAGACC
TNF-α	GTGACAAGCCTGTAGCCCAT	CTCTGATGGCACCACCAACT
SMAD6	AAAACCGTCACGTACTCGCT	GGTCGTACACCGCATAGAGG

2.4. Animal Model

A double blind, split-mouth design, randomized controlled trial with sheep was designed and reported according to the ARRIVE (Animal Research: Reporting of In-vivo Experiments) and CONSORT (CONsolidated Standards of Reporting Trials) guidelines. The PICOT question was as follows: Did the sheep (P) receiving revitalisation therapy with Biodentine (I) in comparison to those receiving ProRoot WMTA (C) show a different histological outcome (O) after 6 months' follow-up (T)? Due to lack of previous literature with quantitative histological data upon which sample size analysis could be calculated, we decided to perform a pilot study with post hoc sample size analysis. In accordance to a study by Altaii et al. [17] with a similar model (sheep), therapy (regenerative therapy with TCS cement) and protocol, an initial sample size of four teeth per treatment group was considered for this pilot study. The study protocol was approved by the ethical committee on animal experiments, Faculty of Medicine and Health Sciences, University Hospital, Ghent, Belgium (ECD 16/40). Four Suffolk sheep (2 males and 2 females) aged between 12 and 18 months with two newly erupted mandibular immature central incisors (two-tooth stage) were recruited. The sheep were housed in pairs in a stable with natural light during the treatment period. A preoperative digital occlusal radiograph (Dürr Dental Ag and VistaScan Perio, Bietigheim-Bissingen, Germany) was taken to ensure the presence of an immature open apex.

2.5. Randomization and Blinding

Simple randomization was performed for both the tooth and the biomaterial group depending on a set of randomized numbers generated with a computer using Matlab version 8.0 (https://nl.mathworks.com/products/matlab.html, accessed on 10 July 2021, The Mathworks Inc., Natick, MA, USA) software. The randomization list was concealed in a sealed envelope. The chosen tooth was randomly allocated by an independent individual (S.R) to one of the two groups: the Biodentine (Septodont, Saint Maur des Faussés, France) group or the ProRoot WMTA (Dentsply, Tulsa Dental, OK, USA) group. The other tooth was automatically allocated to the other group. All teeth were coded with an eight-digit code, where the first four digits were coded for the sheep number and the last four digits were coded for the tooth number and the biomaterial used. These unique codes ensured that the assessors (oral pathologists) were blinded to the type of treatment performed. A single precalibrated well-experienced operator (L.M) performed all the intervention and was blinded to the treatment group until the placement of the biomaterial.

2.6. General Anaesthesia

All procedures were performed under general anaesthesia. First, premedication (Domosedan, detomidine hydrochloride, 0.04 mg/kg intramuscular) was administered to the sheep. Induction of anaesthesia was implemented intravenously with midazolam 0.2 mg/kg and ketamine 6 mg/kg. Anaesthesia was maintained using 2% sevoflurane. Duratears, (Dextran 70, Hypromellose) was used for the protection of the cornea due to risk of dehydration.

2.7. Intervention

Endodontic revitalisation therapy was conducted according to the European Society of Endodontology (ESE) position statement on revitalisation procedures [33].

2.8. Phase 1

Coronal access was obtained lingually with a high-speed cylindrical diamond bur, and the pulp tissue was inoculated with a mixture of supragingival plaque and 2 mL of saline (Mini-Plasco, Braun, Diegem, Belgium). A cotton pellet soaked in the plaque suspension was placed on the exposed pulp tissue, and the access cavity was left open to induce necrosis of the pulp. All sheep were housed under the same conditions, fed in line with a natural diet and monitored for symptoms of postoperative pain for 4 weeks.

2.9. Phase 2

At each phase, the teeth were clinically examined for fractures, gingival swelling, abscess and/or fistula. Experimental teeth were isolated with a rubber dam, and the access cavity was reopened with a high-speed cylindrical diamond bur. Necrosis was confirmed clinically by the presence of necrotic pulp tissue and/or intracanal pus. At this stage, in all the sheep, a thin dentine bridge was observed both clinically and radiographically at the middle third of the root. The thin dentine bridge was mechanically removed with endodontic files. For reasons of anatomy, tooth orientation in the sheep mandible and facilitation of instrumentation, the access cavity was enlarged into the buccal area of the crown. After removal of all loose/necrotic pulp tissue, mild mechanical instrumentation of the root canal was performed. Root canals were irrigated with 20 mL of 3% sodium hypochlorite (NaOCl, Denteck, Zoetemeer, Belgium) using a side-vented needle. Bleeding or exudate within the root canal was controlled with paper points and irrigated with 5 mL of sterile physiological saline. After a drying with paper points (Dentsply, York, PA, USA), the root canal was finally irrigated with 20 mL of 17% ethylenediamenetetraacetic acid (EDTA, Denteck, Zoetemeer, Belgium). Calcium hydroxide paste (Ultracal, Ultradent, UT, USA) was injected in the root canal, a sterile cotton pellet was placed on the intracanal dressing and the cavity was sealed with glass-ionomer cement (Ketac Fill Aplicap; 3M ESPE, Seefeld, Germany). All sheep were housed under the same conditions, fed in line with a natural diet and monitored for symptoms of postoperative pain for 2 weeks. Discomfort in sheep was evaluated initially by means of monitoring behavioural changes such as pacing, agitation, vocalization and decreased appetite. Upon identification of behavioural changes, physiological parameters such as pupil dilatation and change in respiratory rate were investigated.

2.10. Phase 3

The access cavity was reopened with a high-speed cylindrical diamond bur, and the intracanal dressing was removed using irrigation with sterile physiological saline. After a drying with paper points, the root canal was irrigated with 20 mL of 17% EDTA followed by 5 mL of sterile physiological saline. After a drying with paper points, a sterile size-40 Hedström file (Dentsply, Washington, DC, USA) was used to mechanically irritate the apex in a rotational movement to induce bleeding into the root canal. The canal was allowed to fill with blood up to 2 mm below the gingival margin. Hemocollagene (Septodont, Saint Maur des Faussés, France) was placed at the coronal part of the root canal to a height of 2–3 mm on top of the blood clot.

At this stage, the tooth was allocated to one of the intervention groups. In both the groups, the cement was mixed according to the manufacturer's instructions, and a homogenous layer of about 2 mm was placed underneath the cemento–enamel junction. The tooth was restored with glass-ionomer cement (Ketac Fill Aplicap; 3M ESPE, Seefeld, Germany). The steps of phase 3 are shown in Figure 1. During the follow-up period of 6 months, the sheep were maintained and fed in their natural habitat.

A follow-up digital occlusal radiograph (Dürr Dental Ag and VistaScan Perio, Bietigheim-Bissingen, Germany) was taken, and the teeth were extracted. The extracted teeth were stored in 10% neutral-buffered formalin (PFA, VWR International, Radnor, PA, USA). After extraction of the teeth, the sheep were followed up for one week and sent to rehabilitation.

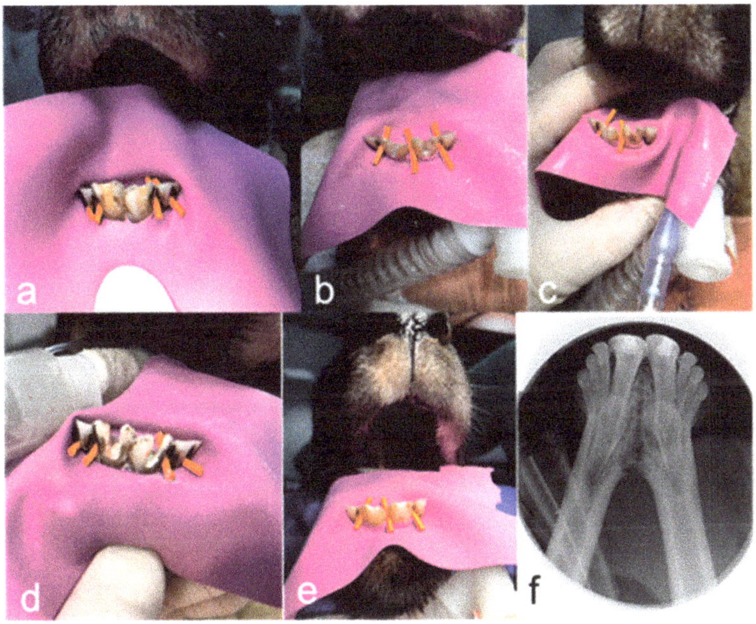

Figure 1. Representative clinical and radiographic images. (**a**) Preoperative image after rubber dam isolation of teeth 31 and 41. (**b**) Visualization of calcium hydroxide after access opening. (**c**) Induction of bleeding in the root canal of tooth 41. (**d**) Placement of intracoronal sealing material: Biodentine in tooth 31 and ProRoot White MTA in tooth 41. (**e**) Restorative build-up of teeth 31 and 41 after revitalisation therapy. (**f**) Postoperative radiographic image of teeth 31 and 41. Phase 4.

2.11. Radiographic Analysis

Quantitative radiographic analysis of intraoral radiographs was not possible, as the radiographs could not be standardized. Cone-beam μCT scans were acquired on a small animal radiation research platform (SARRP, XStrahl, Surrey, UK) [34]. A tube voltage of 90 kV was used in combination with a tube current of 200 μA and a 1 mm Al-filter. A total of 2000 projections were acquired over 360 degrees on a 20 × 20 cm flat-panel amorphous silicon detector with 1024 × 1024 pixels using a circular trajectory with continuous rotation and a magnification factor of 1.5. The acquired projection data were reconstructed into a three-dimensional DICOM image with a 512 × 512 × 512 matrix and a 75 μm isotropic voxel. Radiographic measurements of total root canal area and area of mineralized tissue were performed by setting binary thresholds using Digimizer software version 5.4.1 (https://www.digimizer.com, accessed on 10 July 2021, MedCalc Software, 8400 Ostend, Belgium) on 10 central slices per tooth. Percentage of mineralized tissue within the root canal was calculated as the ratio of the area of mineralized tissue to that of the total area of the root canal.

2.12. Histological Analysis

The teeth were decalcified in a combination of hydrochloric acid and EDTA (Decalcifier DC2 QPath, VWR International, Oud Heverlee, Belgium) and were embedded in paraffin wax after complete decalcification. Serial longitudinal sections 5 μm in thickness were cut and stained with haematoxylin and eosin (VWR International, Oud Heverlee, Belgium) for histological analysis. All slices were scanned with a microscope (Olympus BX51, Olympus, Tokyo, Japan) equipped with Xcellence software version 2.7 (https://www.olympus-lifescience.com/en, accessed on 10 July 2021, Olympus, Tokyo, Japan). All images were analysed independently by an oral pathologist (E.R) and an expert in oral biology (P.D.C).

Three continuous central sections with well-preserved tissue and absence of any artefacts were selected for histological analyses. For each section, 15 to 20 representative fields at 200× magnification covering the entire section were analysed. The following histopathological findings were evaluated using Digimizer software version 5.4.1 (https://www.digimizer.com, accessed on 10 July 2021, MedCalc Software, 8400 Ostend, Belgium).

1. Extent of inflammation was scored from 0 to 4 as follows: score 0, absent—absence of inflammatory cells; score 1, mild—small number of scattered inflammatory cells; score 2, moderate—some foci of inflammatory cells; score 3, severe—intense infiltration with inflammatory cells and altered tissue architecture; and score 4: necrosis: amorphous clumps of tissue remnants.
2. Presence or absence of tissue with cellularity and vascularity inside the pulp space was scored from 0 to 3 as follows: score 0—no tissue in-growth into the canal space; score 1—evidence of tissue in-growth into the apical third of the canal; score 2—evidence of tissue in-growth extending to the middle third of the canal; and score 3—evidence of tissue in-growth extending to the cervical third of the canal.
3. For area of tissue with cellularity and vascularity, only soft tissue with the presence of cells and blood vessels were measured.
4. Length of odontoblast lining attached to the dentinal wall was measured on both sides of the root for each tooth.
5. Number of blood vessels in each section was counted.
6. Area of blood vessels expressed as percentage of vascularity was calculated as the ratio of area of blood vessels to that of the area of tissue with cellularity and vascularity within each histological section.
7. The area of empty root canal space was measured as the area inside the root canal where neither soft nor hard tissue structures were present.

The ordinal scores for the first two parameters were adapted from previous studies by Tawfik et al. [35]. and Fahmy et al. [36]. The other parameters were measured on a continuous numerical scale to allow effect size and power analysis calculations.

2.13. Statistical Analysis

All measurements were performed by two independent evaluators (E.R and S.R), and any discrepancy in scoring was solved by a third person (P.DC). The evaluators repeated the measurements after two weeks. The interclass correlation coefficient and intraclass correlation coefficient were calculated using Statistical Package for Social Sciences (SPSS) v25.0 (IBM Corp., Armonk, NY, USA). All continuous data were subjected to statistical analysis with the Wilcoxon matched-pairs signed rank test at a significance level of $p < 0.05$ with GraphPad Prism version 6 (https://www.graphpad.com, accessed on 10 July 2021, GraphPad software Inc., San Diego, CA, USA). Effect size (Hedges', g) was calculated for all continuous measurements. Cohen's effect size index was used to classify the effect size as small (0.2), medium (0.5), large (0.8) and very large (1.3) [37]. GPower version 3.1 [38] (http://www.psychologie.hhu.de/arbeitsgruppen/allgemeine-psychologie-und-arbeitspsychologie/gpower.html, accessed on 10 July 2021) was used to perform calculations on sample size, effect size and statistical power. The minimal significance (α) and statistical power ($1 - \beta$) were set at 0.05 and 0.80, respectively.

3. Results

3.1. Gene Expression

Differential gene expression was observed between the two TCS-based biomaterials on eight specific gene markers (*TGF-β*, *BMP2*, *BGLAP*, *VEGFA*, *WNT5A*, *MMP1*, *TNF-α* and *SMAD6*) known to play a role in the mineralisation, angiogenesis and osteo/odontoblastic differentiation of stem cells (Figure 2). All genes were upregulated by both Biodentine and ProRoot WMTA, but *TNF-α* was not expressed by ProRoot WMTA. With the exception of

BGLAP, a larger increase in mRNA expression (fold change) was observed with Biodentine compared to ProRoot WMTA after 1 day.

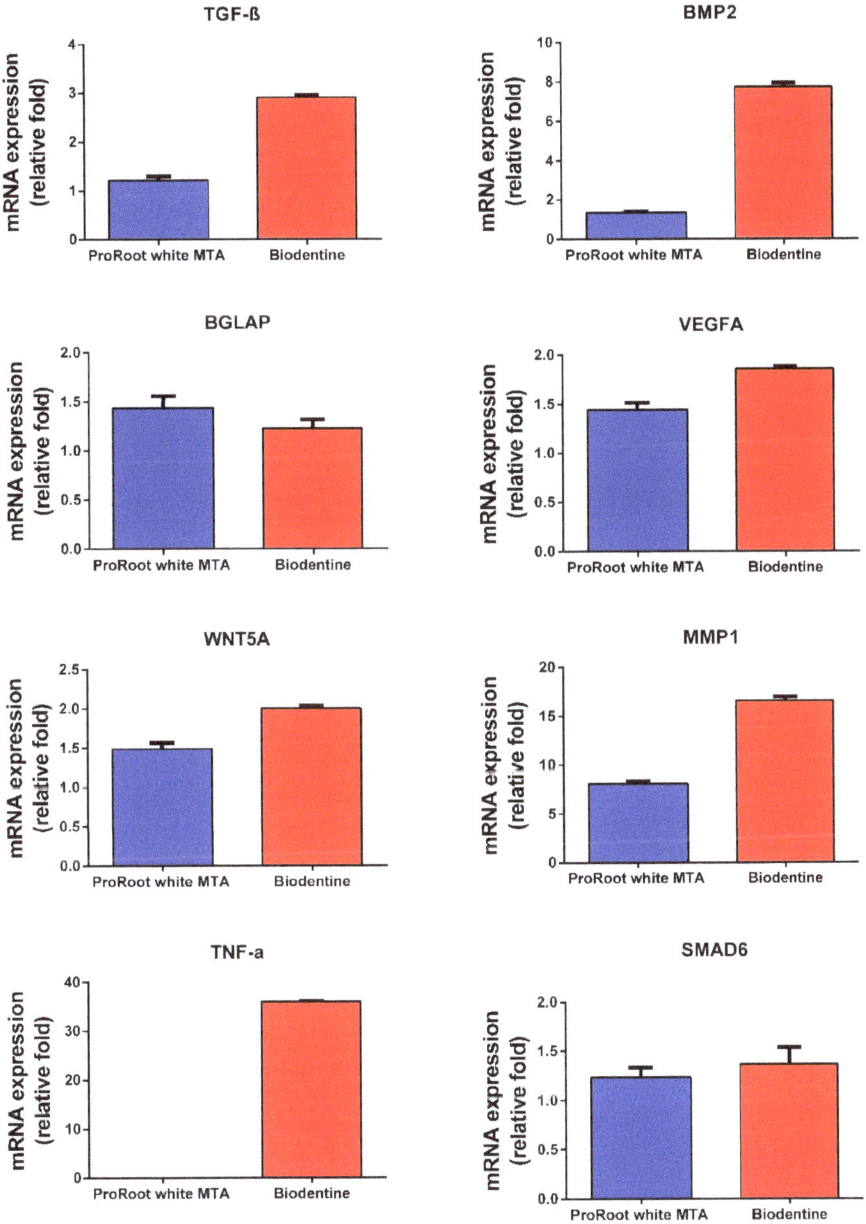

Figure 2. qRT-PCR expression. mRNA expression of TGF-β, BMP2, BGLAP, VEGFA, WNT5A, MMP1, TNF-α and SMAD6 after 1 day. The histogram shows upregulated mRNA expression in relative fold change. The details of the specific primers used for gene expression analysis are provided in Table 1.

3.2. Qualitative Analysis of Revitalisation Therapy

None of the sheep had any discomfort after the revitalisation treatment. One sheep showed symptoms of pain between phase 2 and phase 3 (after the placement of the intracanal dressing). Buprenorphine 0.01 mg/kg was administered intramuscularly, and the pain subsided after three days. After phase 3 treatment, the animals were housed in their natural habitat for a period of 6 months. During this period, one sheep experienced an avulsion of one tooth (treated with Biodentine). The other tooth treated with ProRoot WMTA in the same sheep was intact and was included in the analysis. No other adverse events were observed during the trial. The radiographic outcomes were in accordance with the histological analyses.

Teeth treated with Biodentine (n = 3) presented 2 distinctive histological patterns of the newly regenerated tissue. Two of the three teeth treated with Biodentine (Figure 3) showed clear deposition of tertiary dentine attached to the dentinal walls resulting in root wall thickening. The apices showed lengthening of the root and narrowing by cementum-like tissue. Periodontal ligaments surrounding the teeth were dense and vascular connective tissue was seen. Middle and apical root thirds showed signs of pulp vitality with odontoblast-like cells lining the root canal. The canal showed a rich neoproliferation of blood vessels, nerves and other cells such as fibroblasts. In one of these teeth, obliteration at the coronal third of the pulp was observed.

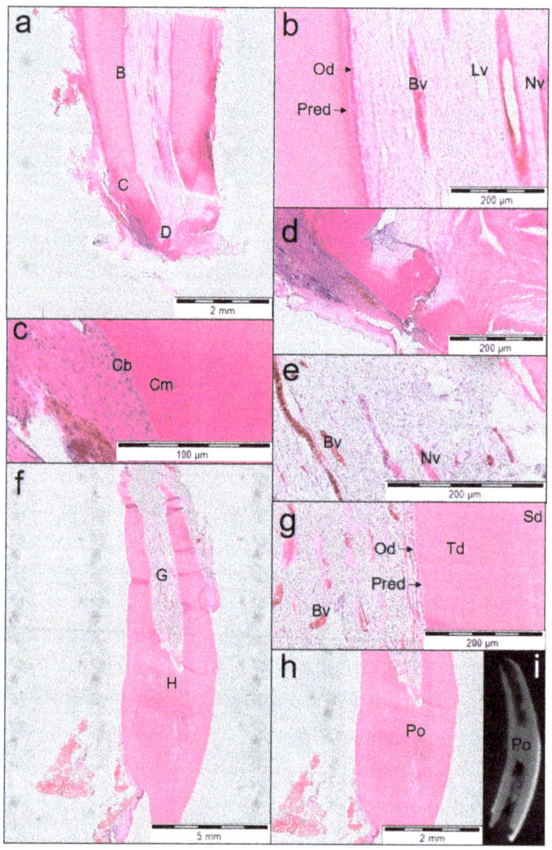

Figure 3. Overview of the first and second teeth treated with Biodentine (haematoxylin and eosin staining; (**a–e**) the first tooth; (**f–i**) the second tooth). (**a**) Midsagittal section of the middle and the apical

part of the root (first tooth). Individual letters (B, C, D) represent the regions of interest. (**b**) Magnified view of region B from (**a**) shows odontoblast (Od) lining attached to the dentinal walls and neoproliferation of blood vessels (Bv) with a dense amount of collagen fibres which is composed of a large number of fibroblasts maintained throughout the middle and apical third of the root. Predentine (Pred), lymph vessels (Lv) and nerves (Nv) can also be seen in this section. (**c**) Magnified view of the region C from (**a**) shows cementum (Cm) with a lining of cementoblast-like cells (Cb). (**d**) Magnified view of the apical region D from (**a**) showing cellular cementum on both sides with narrowing of the apical foramen. (**e**) Section of the first tooth showing nerves (Nv) and blood vessels (Bv) in the middle third of the root. (**f**) Midsagittal section of the coronal, middle and apical third of the root (2nd tooth). Individual alphabets (G, H) represent regions of interest. (**g**) Magnified view of the region G from (**f**) showing reorganized pulp-like tissue in the middle and apical third of the tooth. Excessive neoproliferation of blood vessels (Bv) and restructured pulp-like tissue with odontoblast-like cells lining the dentinal walls. Secondary dentine (Sd), tertiary dentine (Td) and predentine (Pred) with odontoblast (Od) lining. (**h**) Magnified view of the coronal third of the pulp (region H from (**f**)) showing coronal pulp obliteration (Po). (**i**) µCT image showing coronal pulp obliteration (Po).

One tooth treated with Biodentine (Figure 4) showed closing of the apex by cementum-like tissue even though periapical inflammation was present. This tooth also showed a distinguished layer of tertiary dentine attached to the dentinal walls and a dentinal plug separating vital from necrotic pulp tissue. The middle and the apical third of the canal showed fibrous connective tissue with poor cellularity.

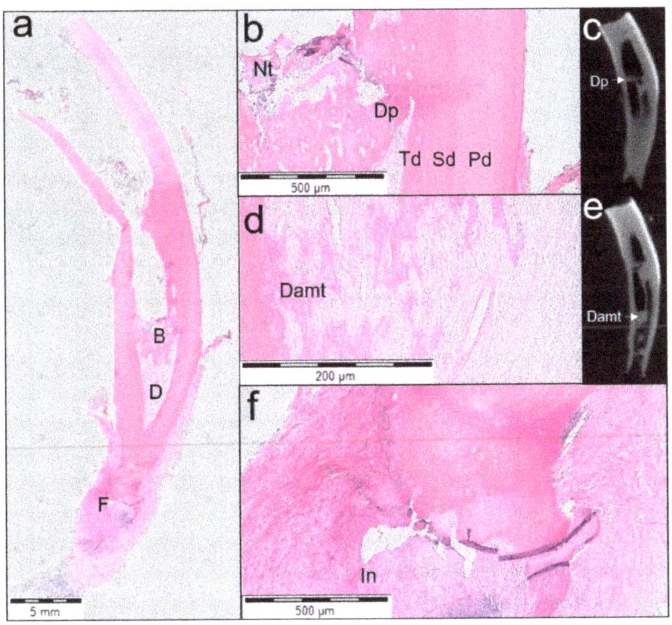

Figure 4. Overview of the third tooth treated with Biodentine (haematoxylin and eosin staining). (**a**) Midsagittal section of coronal, middle and apical third of the root (3rd tooth). Individual letters (B, D, F) represent regions of interest. (**b**) Magnified view of region B from (**a**) shows three distinguished layers of dentine (primary dentine (Pd), secondary dentine (Sd) and tertiary dentine (Td) with a dentine plug (Dp) separating the necrotic tissue (Nt). (**c**) A µCT image showing the dentine plug (Dp). (**d**) Magnified view of region D from (**a**) shows dentine-associated mineralisation (Damt) with fibrous connective tissue. (**e**) A µCT image showing dentine-associated mineralisation tissue (Damt). (**f**) Magnified view of the periapical region (region F from (**a**)) showing chronic inflammation with rich inflammatory infiltration (In) and apex closure.

Teeth treated with ProRoot WMTA also presented two distinctive histological patterns of the newly regenerated tissue. One tooth treated with ProRoot WMTA (Figure 5) presented with a distinguished layer of tertiary dentine which was attached to the dentinal walls on only one side and detached on the other side. Lengthening of the roots with cementum-like tissue and a discontinuous lining of odontoblast-like cells in the apical third of the root were seen. This was observed only unilaterally; very few odontoblast-like cells were seen at the contralateral side. The middle and the coronal third of the roots were obliterated with osteodentine-like structures with dispersed cellular inclusions. The middle and the apical third of the canal spaces contained fibrovascular connective tissue with poor cellularity.

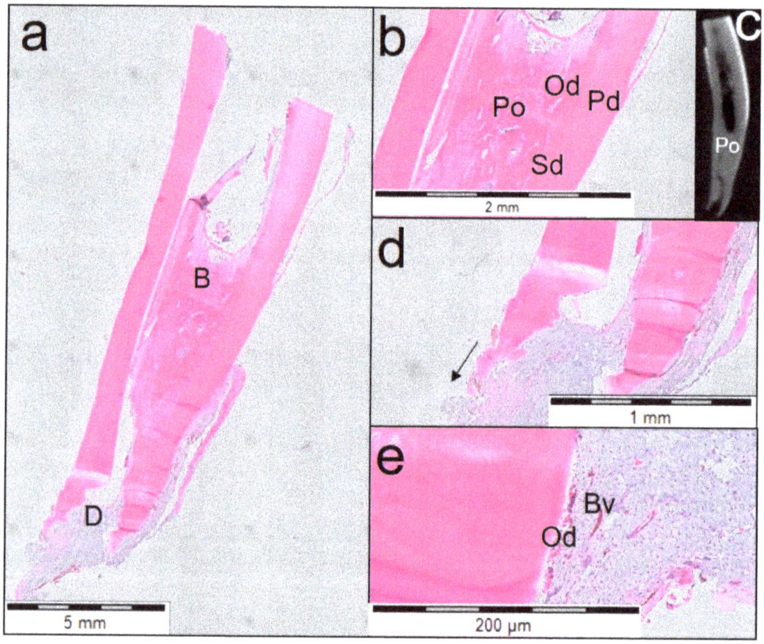

Figure 5. Overview of the first tooth treated with ProRoot WMTA (haematoxylin and eosin staining). (**a**) Midsagittal section of the coronal, middle and apical third of the root (first tooth). Individual letters (B, D) represent regions of interest. (**b**) Magnified view of region B from (**a**) shows pulp obliteration (Po) by osteodentine (Od), and the dentinal wall on the right shows complete void throughout the canal. Primary dentine (Pd) and secondary dentine (Sd) can also be seen in this section. (**c**) A μCT image showing pulp obliteration (Po). (**d**) Magnified view of region D from (**a**) shows lengthening of the apical third of the root (indicated by an arrow). (**e**) Magnified view of the apical third of the root showing few areas with discontinuous lining of odontoblast-like cells (Od) and blood vessels (Bv). Fibrovascular connective tissue with poor cellularity is seen.

In three of the four teeth treated with ProRoot WMTA (Figure 6), no distinguished layer of tertiary dentine attached to the dentinal walls was observed, and the root walls were remarkably thin. The apical and middle thirds of the root canals were filled with loose, sparse fibrotic tissue with areas showing aggregates of red blood cells and inflammatory cells. No prominent narrowing of the root apices was observed. In one tooth, an inflammatory cell infiltrate was observed at the apical region.

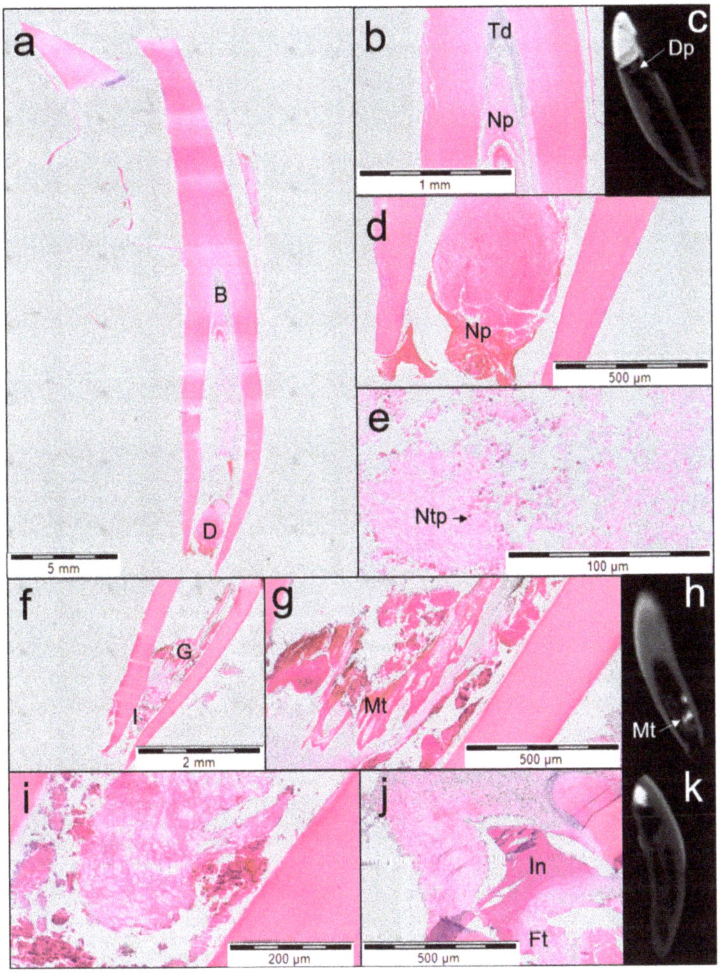

Figure 6. Overview of the second, third and fourth teeth treated with ProRoot WMTA (haematoxylin and eosin staining; (**a–e**) second tooth; (**f–i**) third tooth; (**j**,**k**) fourth tooth). (**a**) Parasagittal section of the coronal, middle and apical third of the root (secibd tooth). Individual letters (B, D) represent regions of interest. (**b**) Magnified view of region B from (**a**) shows moderate formation of tertiary dentine (Td) and necrotic pulp (Np) in the root canal. (**c**) A μCT image showing formation of a dentine plug (Dp) not attached to the dentine walls on the left side (**d**) Magnified view of region D from (**a**) showing the apical third of the canal with open apex, necrotic pulp (Np) with aggregation of red blood cells and few inflammatory cells. (**e**) Magnified view of the middle third of the root shows necrotic tissue with moderate infiltration of neutrophils (Ntp). (**f**) Midsagittal section of the apical and middle third of the root (3rd tooth). Individual letters (G, I) represent regions of interest. (**g**) Magnified view of region G from (**f**) shows mineralized tissue (Mt) not attached to the dentinal wall. (**h**) A μCT image showing mineralized tissue (Mt). (**i**) Magnified view of region I from (**f**) showing loosely arranged fibrotic tissue filling the apical and middle third of the root, with blood cells and few residual inflammatory cells indicating incomplete recovery from inflammation. (**j**) Inflammatory infiltration (In) in the apical third of the root canal and infection with an open apex. Fibrotic tissue (Ft) can be seen between the apical third and the middle third of the root with few inflammatory cells. (**k**) A μCT image showing fibrotic tissue in the root canal.

3.3. Quantitative Analysis of Revitalisation Therapy

The interclass correlation coefficient between the evaluators was excellent (0.961; CI 0.888–0.987). The intraclass correlation coefficient was also excellent: 0.984 (CI 0.943–0.996) for evaluator 1 (E.R) and 0.98 (CI 0.941–0.993) for evaluator 2 (S.R). The results of the histological and radiographic analyses are summarized in Table 2. Significantly more neo-formed intracanal tissue with cellularity and vascularity was seen in the Biodentine group compared to the ProRoot WMTA group ($p < 0.05$). Likewise, the length of odontoblast lining attached to the dentinal wall was significantly increased in teeth treated with Biodentine compared to those treated with ProRoot WMTA ($p < 0.05$). No significant difference between the treatment groups were observed with the other parameters measured.

Table 2. Quantification of histological and radiographic data.

	Biodentine						ProRoot WMTA							
	Tooth 1		Tooth 2		Tooth 3		Tooth 1		Tooth 2		Tooth 3		Tooth 4	
	Mean	SEM	Mean	SEM	Mean	SEM	Mean	SEM	Mean	SEM	Mean	SEM	Mean	SEM
Extent of inflammation (0–4)	**0**	0	**0**	0	**0**	0	**2**	0	**3**	0	**0**	0	**3**	0
Presence or absence of tissue with cellularity and vascularity inside the pulp space (0–3)	**2**	0	**2**	0	**0**	0	**0**	0	**0**	0	**1**	0	**0**	0
Area of tissue with cellularity and vascularity (mm^2)	**13.14**	1.6	**11.78**	1.47	**0**	0	**0**	0	**0**	0	**0.23**	0.12	**0**	0
Length of odontoblast lining (mm)	**13.4**	1.29	**11.27**	1.43	**0**	0	**0**	0	**0**	0	**1.41**	0.72	**0**	0
Number of blood vessels (n)	**64.33**	19.33	**176.33**	19.62	**0**	0	**0**	0	**0**	0	**45**	16.77	**0**	0
Area of blood vessels expressed as percentage of vascularity (%)	**7.45**	1.85	**10.03**	0.99	**0**	0	**0**	0	**0**	0	**3.76**	0.16	**0**	0
Area of empty root canal space (mm^2)	**1.36**	0.51	**1.15**	0.38	**23.70**	1.08	**11.69**	0.37	**12.91**	2.71	**11.03**	0.71	**22.28**	2.32
Area of mineralized tissue (%)	**0.33**	0.01	**0.26**	0.01	**0.28**	0.02	**0.27**	0.01	**0.08**	0.01	**0.38**	0.02	**0.31**	0.01

Hedges' g is a measure of effect size weighted according to the relative size of each sample. It was used in the present study due to the discrepancy in sample size between the Biodentine and ProRoot WMTA groups, as one tooth in the Biodentine group was lost from avulsion during follow-up. In correlation with the significant results obtained, effect size could be categorized as very large for both areas of tissue with cellularity and vascularity (1.92) and length of odontoblast lining attached to the dentinal wall (1.86). For the number of blood vessels and percentage of vascularity, the effect size was large (1.46 and 1.27, respectively), but power analysis revealed inadequate power (0.26 and 0.32, respectively). Post hoc power analysis estimated that a sample size of a minimum of six teeth per group was necessary to achieve a power of 0.80. The effect size for the area of empty root canal space was medium (0.69), resulting in inadequate power (0.12), and a sample size no smaller than 36 teeth per group would be required to achieve a power of 0.80. The percentage of mineralized tissue within the root canal showed a small effect size (0.31) with inadequate power (0.06). A large sample size of at least 169 teeth per group would be required to achieve an adequate power (0.80) to show significant difference in the percentage of mineralized tissue.

The columns are colour coded such that the columns with same colour indicate the teeth from a single sheep. The data for histological quantification is calculated as the average from n = 3 sections for each tooth and from n = 10 slices for radiographic data. SEM values represent the standard error of the mean. Extent of inflammation was scored from 0–4. Score 0: Absent: Absence of inflammatory cells; Score 1: Mild: Small number of scattered inflammatory cells; Score 2: Moderate: Some foci of inflammatory cells; Score 3: Severe: Intense infiltration with inflammatory cells and altered tissue architecture; Score 4: Necrosis: Amorphous clumps of tissue remnants. Presence or absence of tissue with cellularity and vascularity inside the pulp space was scored from 0–3. Score 0: No tissue in-growth into the canal space; Score 1: Evidence of tissue in-growth into the apical third of the canal; Score 2: Evidence of tissue in-growth extending to the middle third of the canal; Score 3: Evidence of tissue in-growth extending to the cervical third of the canal. Area of blood vessels expressed as percentage of vascularity was calculated as the ratio of area of blood vessels to that of the area of tissue with cellularity and vascularity within each histological section. Area of mineralized tissue was expressed in percentage calculated as the ratio of area of mineralization to that of the total area of the root canal.

4. Discussion

Tri- and dicalcium silicate, tricalcium aluminate, calcium sulfate dihydrate and bismuth oxide constitute the powder component of ProRoot WMTA. On the other hand, the powder component of Biodentine is composed of tricalcium silicate, calcium carbonate, calcium oxide, iron oxide and zirconium oxide. The powder component of ProRoot WMTA is mixed with distilled water, but Biodentine is mixed with a liquid containing hydrosoluble polymer and calcium chloride to accelerate the setting time [39]. Although ProRoot WMTA and Biodentine are classified as TCS-based hydraulic cements, there is difference in the nature of synthesis, additives added, size of the particles and the composition of liquid medium [40]. These differences in composition is reflected in the difference in physical (compressive strength, push-out bond strength, density and porosity), biologic (immediate formation of calcium hydroxide, higher release and depth of incorporation of calcium ions) and clinical properties (handling and tooth discoloration) [41].

For studying dental pulp revitalisation outcomes, diphyodont animals such as ferret, rabbit, dog, sheep and primates with immature single-rooted permanent teeth are the model of choice [42]. Recent literature supports the use of sheep in the two-tooth stage as an appropriate animal model for revitalisation research due to their anatomic similarity to human teeth [17,43]. A split-mouth design was devised to eliminate the intersheep variability in terms of age of the sheep, developmental stage of the tooth and individual immune response from affecting the outcome of revitalisation therapy. In the current protocol, 3% NaOCl was chosen for disinfection and dissolution of pulp tissue remnants to limit the toxicity towards stem cells [44]. The concentration of NaOCl recommended by the ESE and American Association of Endodontologists (AAE) guidelines for revitalisation lies in the range of 1.5–3% to achieve the balance between sufficient disinfection and tissue preservation. EDTA was used, as it is the preferred choice of irrigant in the revitalisation protocol based on its ability to inhibit biofilm formation and demineralize the dentine to release growth factors sequestered into the calcified dentine matrix [5,45]. Calcium hydroxide was the intracanal medicament of choice, as antibiotic pastes can cause discoloration, sensitization, resistance and difficulty of removal from the root canal [46,47]. Both the ESE and AAE guidelines for revitalisation recommend calcium hydroxide as the first choice for disinfection, and the use of antibiotic pastes is recommended only in cases of persistent infection. Moreover, calcium hydroxide can prompt mineralisation due to its ability to solubilize dentine extracellular matrix and release growth factors [48]. Hemocollagene was verified to be beneficial for the placement of the TSCs at their optimum level without apical displacement or distortion of the newly formed blood clot [49].

In this study, blood clot was used as a scaffold. It stimulates the proliferation and differentiation of fibroblasts, odontoblasts and cementoblasts from their undifferentiated mesenchymal stem cells via platelet-derived growth factor, vascular endothelial growth factor and other tissue growth factors present in the blood clot. However, the disadvantages of using blood clot as a scaffold may include the uncertain composition, unknown breakdown kinetics, inability to have adequate bleeding in all cases and the necessity to traumatize apical tissues for evoking bleeding [50].

During the placement of intracanal medication (phase 2), a dentine bridge was clinically and radiographically observed in all teeth in the middle third of the roots. The inoculation with supragingival plaque (phase 1) mimicked the advancing infection found in deep (dentinal) caries lesions, where an acute immune reaction is triggered resulting in the formation of a dentine bridge [51]. During phase 2 pulpectomy, the pulp tissue proximal to the dentine bridges was clinically found to be necrotic, indicating that the initial immune reaction had failed to stop the propagation of local necrosis started at the invasion front. Previous authors indicated that, amongst others, the size of pulp exposure and the presence of infection are critical elements in success or failure of pulp repair [52,53].

Galler et al. characterized regeneration seen following revitalisation therapy as the restoration of pulp tissue architecture and function. Histologically, pulp-like tissue formation in root canals is observed where stem cells differentiate into odontoblasts. Conversely, repair has been defined as ectopic tissue formation with partial loss of function and formation of fibrous tissue, cementum or bone inside the root canal [33]. If this categorization was to the results of the present study, two of the three teeth treated with Biodentine showed regeneration while one tooth exhibited repair. On the other hand, histological outcomes of the ProRoot WMTA group revealed regeneration in one tooth and repair in three of the four treated teeth. However, the intracoronal sealing materials revealed significantly different histological outcomes; clinically and radiographically, all treated teeth were asymptomatic. By following the ESE guidelines for revitalisation of immature carious necrotic teeth, clinical and radiographic success was possible, while the histological outcome was unpredictable and nonreproducible.

A majority of the previous literature on regenerative endodontics fails to associate newly formed calcified tissues in the pulp space with the presence of odontoblasts [17]. Differentiation of stem cells into odontoblasts is a prerequisite for the formation of pulp-like tissue in root canals [54]. Biodentine and ProRoot WMTA are known to activate extracellular signal-regulated kinase $\frac{1}{2}$, nuclear factor E2-related factor 2, p38, c-Jun N-terminal kinase mitogen-activated protein kinase, p42/p44 mitogen-activated protein kinase, nuclear factor kappa B and fibroblast growth factor receptor pathways to stimulate the osteogenic/odontogenic capacity of dental pulp stem cells via proliferation, angiogenesis and biomineralisation. The upregulated expression of *BMP2* indicates the ability of these cements to induce differentiation of dental pulp stem/progenitor cells into odontoblasts [55]. Our results show that Biodentine induced a fourfold higher expression of BMP2 compared to ProRoot WMTA. The upregulation of *SMAD6* limits *BMP* signalling for proper odontoblastic differentiation, as *SMAD6* provides feedback inhibition of BMP-receptor activation [56]. *BGLAP* is an osteoblast marker and is also considered to be a late differentiation marker of odontoblasts [57]. Both Biodentine and ProRoot WMTA upregulated the expression of *SMAD6* and *BGLAP*. In the present study, two teeth treated with Biodentine showed signs of a vital pulp with odontoblast-like cells lining the middle and apical third of the canal. In one of the teeth treated with ProRoot WMTA, the root canal was continuous with the periodontal ligament indicating a lack of closure of the apex. In these teeth, odontoblast-like cells were seen in a few zones which may hint at a probability of complete pulpal regeneration in the future.

Two teeth in Biodentine group presented with rich vascularity, lymphatic vessels and few nerves which were absent in the ProRoot WMTA group. Despite the higher number of blood vessels and increased vascularity percentage of Biodentine compared to ProRoot WMTA, no significant differences could be seen between the two groups due to low sample size of this pilot study. Previous authors have corroborated that dental pulp stem cells produce angiogenic and neurotropic factors for revascularization and reinnervation of the regenerated pulp [58–60]. This is in correlation with the results of other studies where the regenerated pulp was found to have high vascularity [61–64] and innervated with newly regenerated nerve fibres [9,61,62]. *VEGF* and *MMP1* are known to promote blood vessel formation-enhancing neovascularization in vivo [65]. Both these genes were upregulated by Biodentine and ProRoot WMTA after 1 day.

Lengthening of the root with cementum-like tissue and three distinguished layers of dentine was seen in all Biodentine-treated teeth and one tooth treated with ProRoot WMTA. Newly deposited dentine-like tissue with entrapped cells inside the calcified matrix was a finding similar to that of previous studies [66,67]. The deposition of tertiary dentine by odontoblasts suggests the differentiation of migrant stem cells elicited by transforming growth factor (*TGF-β*) after revitalisation therapy [68,69]. In the present study, Biodentine showed a higher expression of *TGF-β* than did ProRoot WMTA. Biodentine has been shown to be responsible for early reparative dentine synthesis by inducing *TGF-β1* release from dental pulp stem cells [70]. Similar results with lengthening of root by cementum-like tissue were reported earlier in immature sheep teeth treated with MTA [17]. Outward flow of dentinal fluids following pulp necrosis may cause odontoblast entrapment resulting in tubular or vacuolated reparative dentine [71].

Pulp injury causes secretion of neuropeptides by sensory nerves resulting in neurogenic inflammation. The pulp tissue becomes edematous with a rise in blood and interstitial fluid within the root canal causing the compression of thin-walled venules and increased resistance to blood flow. Decrease in blood flow causes aggregation of red blood cells and the elevation of blood viscosity inducing hypoxia or ischemia, suppressing cellular metabolism in the affected area and finally necrosis of the pulp [71]. This could explain the presence of red blood cells in three of the four ProRoot WMTA-treated teeth showing sparse fibrotic tissue in the canal space and no prominent narrowing of the root apex.

The ability of Biodentine and ProRoot WMTA to upregulate mineralisation-related genes [18], thereby inducing mineralisation and reparative dentine formation [72] has broadened the clinical indications of these materials to direct/indirect pulp capping [73], pulpotomy [74], furcation repair [75,76] and the treatment of resorption [77]. Other authors already have reported that reparative dentine formed by Biodentine in pulp capping studies is usually faster and thicker than are the ProRoot WMTA homologues [78–81]. Our results showed that both Biodentine and ProRoot WMTA upregulated the expression of *WNT5A*, but only Biodentine induced *TNF-α* expression. *WNT5A* and *TNF-α* induce mineralisation and mineralisation-related gene expression through nuclear factor kappa (NF-κB) signalling pathway in hDPSCs [82,83].

Although both Biodentine and ProRoot WMTA are tricalcium silicate-based cements, there existed marked differences in the gene expression profiles of these cements which could be attributed to the difference in purity, composition, particle size, mechanical properties and kinetics of calcium release [40]. Therefore, we hypothesize that the difference in gene expression profiles could be responsible for the different histological outcomes observed in the present study.

The limited number of teeth treated in each group prevented this pilot study from having adequate power ($1 - \beta > 0.80$) and effect size to determine a significant difference in vascularity (number of blood vessels and relative area). Post hoc power analysis showed that the low effect size demanded a very large sample size for area of empty root canal space (n = 36 teeth) and percentage of mineralized tissue in the root canal (n = 169 teeth). However, we believe that these are secondary parameters, as the primary goal of revitalisation therapy is to achieve pulp tissue architecture and function. Thus, the presence of neoformed soft tissue with blood vessels and an odontoblast lining attached to the dentinal wall should be considered as the primary characteristics for evaluating success. Based on the above assumption and taking into account the practical as well as the ethical aspects of conducting an animal study, we recommend that a sample size of a minimum of six teeth per group will be required to achieve adequate power ($1 - \beta > 0.80$) in future studies.

4.1. Limitations of the Present Study

Quantitative radiographic analysis of intraoral radiographs was not possible, as they could not be standardized. Superimposition of radiographs was not feasible owing to the dynamic alterations in the dentition of the sheep, namely, change in coronal height of erupting incisors, continuous attrition of the molars and the shedding of deciduous laterals/canines during the follow-up period. Extending the follow-up period was not possible in the present study due to the high cost of sheep maintenance. The histological outcomes of the present study are a simulation of revitalisation therapy after pulp necrosis subsequent to bacterial exposure, but the same may not be true for trauma or longstanding periapical infection due to possible changes in the viability of Hertwig's epithelial root sheath. The difference in biological dynamics between sheep and humans should be considered before the extrapolation of the current results to humans. Although significant differences between the groups exist for some of the measured histological characteristics, the results need to be interpreted with much caution considering the low sample size of the pilot study. Future studies with larger sample sizes and adequate power as estimated by the results of this pilot study would yield more valuable and concrete information regarding the role of intracoronal sealing biomaterial on the histological outcome of endodontic revitalisation.

4.2. Conclusions

Biodentine induced the formation of a significantly larger area of neoformed tissue with cellularity, vascularity and increased length of odontoblast lining attached to the dentine walls compared to ProRoot WMTA ($p < 0.05$). Hence, the null hypothesis that there is no difference in the histologic outcome between revitalisation therapy using Biodentine or ProRoot WMTA is rejected. The intracoronal sealing material appears to play an ancillary role in the type of tissue formed in the root canal after revitalisation therapy, possibly through an interplay between signalling molecules, responsive cell populations and the microenvironment.

Author Contributions: E.R.: conceptualization, histological acquisition, interpretation, analysis and writing of original draft; S.R.: methodology, writing and editing the final draft; H.D.: histological acquisition and editing the final draft; C.V.: radiological acquisition, interpretation and editing the final draft; P.D.C.: histological analysis and interpretation, supervision and editing the final draft; L.M.: clinical treatment of the sheep, supervision and editing the final draft. All authors have read and agreed to the published version of the manuscript.

Funding: This research was funded by Septodont, Saint Maur des Faussés, France.

Institutional Review Board Statement: The study protocol for isolation of human dental pulp stem cells were approved by the Ethical Committee of University hospital, Ghent, Belgium, according to the laws of the ICH Good Clinical Practice (GE11-LM-go-2006/57). The animal study protocol was approved by the ethical committee on animal experiments, Faculty of Medicine and Health Sciences, University Hospital, Ghent, Belgium (ECD 16/40).

Informed Consent Statement: Informed consent was obtained from all subjects before using the extracted teeth for isolation of human dental pulp stem cells.

Data Availability Statement: The data presented in this study are available on request from the corresponding author. The data are not publicly available as this study is part of an ongoing project and only the result sof the pilot study are presented here.

Acknowledgments: The authors would like to thank Maxime Mahu, Karlijn Debusschere, Sabine de Groote and Lynn de Keyzer for their technical assistance with sheep anaesthesia and housing. We would also like to thank Leen Pieters, Toke Thiron and Jakob Van Acker for helping with the preparation, photography of histological specimens and intraoral radiography, respectively. We thank Srinath Govindarajan for his help with the qRT-PCR experiments.

Conflicts of Interest: Rajasekharan holds a postdoctoral fellowship from the Fund for Scientific Research–Flanders (FWO). Martens would like to report research grants and nonfinancial support (i.e., cement materials free of charge) from Septodont, France and Dentsply, Germany. Outside of the submitted work, Martens and Rajasekharan report personal fees (honoraria/lecture fees/educational courses) and research grants from Septodont, France, negotiated with Ghent University. Rathinam, Declercq, Vanhove and De Coster report no competing interests. The funders had no role in the design of the study; in the collection, analyses or interpretation of data; in the writing of the manuscript; or in the decision to publish the results.

References

1. Tan, L.H.; Wang, J.; Yin, S.; Zhu, W.T.; Zhou, G.D.; Cao, Y.L.; Cen, L. Regeneration of dentin-pulp-like tissue using an injectable tissue engineering technique. *Rsc. Adv.* **2015**, *5*, 59723–59737. [CrossRef]
2. Harlamb, S.C. Management of incompletely developed teeth requiring root canal treatment. *Aust. Dent. J.* **2016**, *61* (Suppl. S1), 95–106. [CrossRef] [PubMed]
3. Kahler, B.; Rossi-Fedele, G.; Chugal, N.; Lin, L.M. An Evidence-based Review of the Efficacy of Treatment Approaches for Immature Permanent Teeth with Pulp Necrosis. *J. Endodont.* **2017**, *43*, 1052–1057. [CrossRef] [PubMed]

4. Altaii, M.; Richards, L.; Rossi-Fedele, G. Histological assessment of regenerative endodontic treatment in animal studies with different scaffolds: A systematic review. *Dent. Traumatol.* **2017**, *33*, 235–244. [CrossRef]
5. Nosrat, A.; Kolahdouzan, A.; Hosseini, F.; Mehrizi, E.A.; Verma, P.; Torabinejad, M. Histologic Outcomes of Uninfected Human Immature Teeth Treated with Regenerative Endodontics: 2 Case Reports. *J. Endodont.* **2015**, *41*, 1725–1729. [CrossRef]
6. Albuquerque, M.T.; Valera, M.C.; Nakashima, M.; Nor, J.E.; Bottino, M.C. Tissue-engineering-based strategies for regenerative endodontics. *J. Dent. Res.* **2014**, *93*, 1222–1231. [CrossRef]
7. Del Fabbro, M.; Lolato, A.; Bucchi, C.; Taschieri, S.; Weinstein, R.L. Autologous Platelet Concentrates for Pulp and Dentin Regeneration: A Literature Review of Animal Studies. *J. Endodont.* **2016**, *42*, 250–257. [CrossRef]
8. Gomes-Filho, J.E.; Duarte, P.C.; Ervolino, E.; Mogami Bomfim, S.R.; Xavier Abimussi, C.J.; Mota da Silva Santos, L.; Lodi, C.S.; Penha De Oliveira, S.H.; Dezan, E., Jr.; Cintra, L.T. Histologic characterization of engineered tissues in the canal space of closed-apex teeth with apical periodontitis. *J. Endodont.* **2013**, *39*, 1549–1556. [CrossRef] [PubMed]
9. Lei, L.S.; Chen, Y.M.; Zhou, R.H.; Huang, X.J.; Cai, Z.Y. Histologic and Immunohistochemical Findings of a Human Immature Permanent Tooth with Apical Periodontitis after Regenerative Endodontic Treatment. *J. Endodont.* **2015**, *41*, 1172–1179. [CrossRef] [PubMed]
10. Nosrat, A.; Kolahdouzan, A.; Khatibi, A.H.; Verma, P.; Jamshidi, D.; Nevins, A.J.; Torabinejad, M. Clinical, Radiographic, and Histologic Outcome of Regenerative Endodontic Treatment in Human Teeth Using a Novel Collagen-hydroxyapatite Scaffold. *J. Endodont.* **2019**, *45*, 136–143. [CrossRef]
11. Hilkens, P.; Meschi, N.; Lambrechts, P.; Bronckaers, A.; Lambrichts, I. Dental Stem Cells in Pulp Regeneration: Near Future or Long Road Ahead? *Stem. Cells. Dev.* **2015**, *24*, 1610–1622. [CrossRef]
12. Yang, J.W.; Zhang, Y.F.; Sun, Z.Y.; Song, G.T.; Chen, Z. Dental pulp tissue engineering with bFGF-incorporated silk fibroin scaffolds. *J. Biomater. Appl.* **2015**, *30*, 221–229. [CrossRef]
13. Chisini, L.A.; Conde, M.C.; Alcazar, J.C.; Silva, A.F.; Nor, J.E.; Tarquinio, S.B.; Demarco, F.F. Immunohistochemical Expression of TGF-beta1 and Osteonectin in engineered and $Ca(OH)_2$-repaired human pulp tissues. *Braz. Oral. Res.* **2016**, *30*, e93. [CrossRef]
14. Zhu, L.; Dissanayaka, W.L.; Zhang, C. Dental pulp stem cells overexpressing stromal-derived factor-1alpha and vascular endothelial growth factor in dental pulp regeneration. *Clin. Oral Investig.* **2019**, *23*, 2497–2509. [CrossRef]
15. Hayashi, Y.; Murakami, M.; Kawamura, R.; Ishizaka, R.; Fukuta, O.; Nakashima, M. CXCL14 and MCP1 are potent trophic factors associated with cell migration and angiogenesis leading to higher regenerative potential of dental pulp side population cells. *Stem. Cell. Res.* **2015**, *6*, 111. [CrossRef]
16. Yoo, Y.J.; Lee, W.; Cho, Y.A.; Park, J.C.; Shon, W.J.; Baek, S.H. Effect of conditioned medium from preameloblasts on regenerative cellular differentiation of the immature teeth with necrotic pulp and apical periodontitis. *J. Endodont.* **2014**, *40*, 1355–1361. [CrossRef] [PubMed]
17. Altaii, M.; Cathro, P.; Broberg, M.; Richards, L. Endodontic regeneration and tooth revitalization in immature infected sheep teeth. *Int. Endod. J.* **2017**, *50*, 480–491. [CrossRef] [PubMed]
18. Rathinam, E.; Rajasekharan, S.; Chitturi, R.T.; Martens, L.; De Coster, P. Gene Expression Profiling and Molecular Signaling of Dental Pulp Cells in Response to Tricalcium Silicate Cements: A Systematic Review. *J. Endodont.* **2015**, *41*, 1805–1817. [CrossRef] [PubMed]
19. Rathinam, E.; Rajasekharan, S.; Chitturi, R.T.; Declercq, H.; Martens, L.; De Coster, P. Gene Expression Profiling and Molecular Signaling of Various Cells in Response to Tricalcium Silicate Cements: A Systematic Review. *J. Endodont.* **2016**, *42*, 1713–1725. [CrossRef] [PubMed]
20. Chan, E.K.M.; Desmeules, M.; Cielecki, M.; Dabbagh, B.; dos Santos, B.F. Longitudinal Cohort Study of Regenerative Endodontic Treatment for Immature Necrotic Permanent Teeth. *J. Endodont.* **2017**, *43*, 395–400. [CrossRef]
21. Estefan, B.S.; El Batouty, K.M.; Nagy, M.M.; Diogenes, A. Influence of Age and Apical Diameter on the Success of Endodontic Regeneration Procedures. *J. Endodont.* **2016**, *42*, 1620–1625. [CrossRef] [PubMed]
22. Huang, G.T. Dental pulp and dentin tissue engineering and regeneration: Advancement and challenge. *Front. Biosci.* **2011**, *3*, 788–800. [CrossRef] [PubMed]
23. Bezgin, T.; Yilmaz, A.D.; Celik, B.N.; Kolsuz, M.E.; Sonmez, H. Efficacy of Platelet-rich Plasma as a Scaffold in Regenerative Endodontic Treatment. *J. Endodont.* **2015**, *41*, 36–44. [CrossRef] [PubMed]
24. Torabinejad, M.; Alexander, A.; Vandati, S.A.; Grandhi, A.; Baylink, D.; Shabahang, S. Effect of Residual Dental Pulp Tissue on Regeneration of Dentin-pulp Complex: An In Vivo Investigation. *J. Endodont.* **2018**, *44*, 1796–1801. [CrossRef]
25. Verma, P.; Nosrat, A.; Kim, J.R.; Price, J.B.; Wang, P.; Bair, E.; Xu, H.H.; Fouad, A.F. Effect of Residual Bacteria on the Outcome of Pulp Regeneration In Vivo. *J. Dent. Res.* **2017**, *96*, 100–106. [CrossRef]
26. Zarei, M.; Jafarian, A.H.; Harandi, A.; Javidi, M.; Gharechahi, M. Evaluation of the expression of VIII factor and VEGF in the regeneration of non-vital teeth in dogs using propolis. *Iran. J. Basic Med. Sci.* **2017**, *20*, 172–177. [CrossRef]
27. Fouad, A.F. Microbial Factors and Antimicrobial Strategies in Dental Pulp Regeneration. *J. Endodont.* **2017**, *43*, S46–S50. [CrossRef]
28. Diogenes, A.R.; Ruparel, N.B.; Teixeira, F.B.; Hargreaves, K.M. Translational science in disinfection for regenerative endodontics. *J. Endodont.* **2014**, *40*, S52–S57. [CrossRef]
29. Nagata, J.Y.; Gomes, B.P.F.D.; Lima, T.F.R.; Murakami, L.S.; de Faria, D.E.; Campos, G.R.; de Souza, F.J.; Soares, A.D. Traumatized Immature Teeth Treated with 2 Protocols of Pulp Revascularization. *J. Endodont.* **2014**, *40*, 606–612. [CrossRef]

30. Zhai, Q.; Dong, Z.; Wang, W.; Li, B.; Jin, Y. Dental stem cell and dental tissue regeneration. *Front. Med.* **2019**, *13*, 152–159. [CrossRef]
31. Karamzadeh, R.; Eslaminejad, M.B.; Aflatoonian, R. Isolation, characterization and comparative differentiation of human dental pulp stem cells derived from permanent teeth by using two different methods. *J. Vis. Exp.* **2012**, *69*, e4372. [CrossRef]
32. Drennan, M.B.; Govindarajan, S.; Verheugen, E.; Coquet, J.M.; Staal, J.; McGuire, C.; Taghon, T.; Leclercq, G.; Beyaert, R.; van Loo, G.; et al. NKT sublineage specification and survival requires the ubiquitin-modifying enzyme TNFAIP3/A20. *J. Exp. Med.* **2016**, *213*, 1973–1981. [CrossRef] [PubMed]
33. Galler, K.M.; Krastl, G.; Simon, S.; Van Gorp, G.; Meschi, N.; Vahedi, B.; Lambrechts, P. European Society of Endodontology position statement: Revitalization procedures. *Int. Endod. J.* **2016**, *49*, 717–723. [CrossRef] [PubMed]
34. Bolcaen, J.; Descamps, B.; Deblaere, K.; Boterberg, T.; Hallaert, G.; Van den Broecke, C.; Decrock, E.; Vral, A.; Leybaert, L.; Vanhove, C.; et al. MRI-guided 3D conformal arc micro-irradiation of a F98 glioblastoma rat model using the Small Animal Radiation Research Platform (SARRP). *J. Neurooncol.* **2014**, *120*, 257–266. [CrossRef]
35. Tawfik, H.; Abu-Seida, A.M.; Hashem, A.A.; Nagy, M.M. Regenerative potential following revascularization of immature permanent teeth with necrotic pulps. *Int. Endod. J.* **2013**, *46*, 910–922. [CrossRef]
36. Fahmy, S.H.; Hassanien, E.E.S.; Nagy, M.M.; El Batouty, K.M.; Mekhemar, M.; Fawzy El Sayed, K.; Hassanein, E.H.; Wiltfang, J.; Dorfer, C. Investigation of the regenerative potential of necrotic mature teeth following different revascularisation protocols. *Aust. Endod. J.* **2017**, *43*, 73–82. [CrossRef]
37. Sullivan, G.M.; Feinn, R. Using Effect Size-or Why the P Value Is Not Enough. *J. Grad. Med. Educ.* **2012**, *4*, 279–282. [CrossRef]
38. Faul, F.; Erdfelder, E.; Lang, A.G.; Buchner, A. G*Power 3: A flexible statistical power analysis program for the social, behavioral, and biomedical sciences. *Behav. Res. Methods* **2007**, *39*, 175–191. [CrossRef]
39. Kahler, B.; Chugal, N.; Lin, L.M. Alkaline Materials and Regenerative Endodontics: A Review. *Materials* **2017**, *10*, 1389. [CrossRef]
40. Rajasekharan, S.; Vercruysse, C.; Martens, L.; Verbeeck, R. Effect of Exposed Surface Area, Volume and Environmental pH on the Calcium Ion Release of Three Commercially Available Tricalcium Silicate Based Dental Cements. *Materials* **2018**, *11*, 123. [CrossRef]
41. Rajasekharan, S.; Martens, L.C.; Cauwels, R.G.; Verbeeck, R.M. Biodentine material characteristics and clinical applications: A review of the literature. *Eur. Arch. Paediatr. Dent.* **2014**, *15*, 147–158. [CrossRef]
42. Kaushik, S.N.; Kim, B.; Walma, A.M.; Choi, S.C.; Wu, H.; Mao, J.J.; Jun, H.W.; Cheon, K. Biomimetic microenvironments for regenerative endodontics. *Biomater. Res.* **2016**, *20*, 14. [CrossRef] [PubMed]
43. Altaii, M.; Broberg, M.; Cathro, P.; Richards, L. Standardisation of sheep model for endodontic regeneration/revitalisation research. *Arch. Oral Biol.* **2016**, *65*, 87–94. [CrossRef]
44. Ruangsawasdi, N.; Zehnder, M.; Patcas, R.; Ghayor, C.; Weber, F.E. Regenerative Dentistry: Animal Model for Regenerative Endodontology. *Transfus. Med. Hemoth.* **2016**, *43*, 359–364. [CrossRef]
45. Lin, L.M.; Shimizu, E.; Gibbs, J.L.; Loghin, S.; Ricucci, D. Histologic and Histobacteriologic Observations of Failed Revascularization/Revitalization Therapy: A Case Report. *J. Endodont.* **2014**, *40*, 291–295. [CrossRef] [PubMed]
46. Lenherr, P.; Allgayer, N.; Weiger, R.; Filippi, A.; Attin, T.; Krastl, G. Tooth discoloration induced by endodontic materials: A laboratory study. *Int. Endod. J.* **2012**, *45*, 942–949. [CrossRef] [PubMed]
47. Berkhoff, J.A.; Chen, P.B.; Teixeira, F.B.; Diogenes, A. Evaluation of triple antibiotic paste removal by different irrigation procedures. *J. Endodont.* **2014**, *40*, 1172–1177. [CrossRef]
48. Graham, L.; Cooper, P.R.; Cassidy, N.; Nor, J.E.; Sloan, A.J.; Smith, A.J. The effect of calcium hydroxide on solubilisation of bio-active dentine matrix components. *Biomaterials* **2006**, *27*, 2865–2873. [CrossRef]
49. Jiang, X.; Liu, H.; Peng, C. Clinical and Radiographic Assessment of the Efficacy of a Collagen Membrane in Regenerative Endodontics: A Randomized, Controlled Clinical Trial. *J. Endodont.* **2017**, *43*, 1465–1471. [CrossRef]
50. Palma, P.J.; Ramos, J.C.; Martins, J.B.; Diogenes, A.; Figueiredo, M.H.; Ferreira, P.; Viegas, C.; Santos, J.M. Histologic Evaluation of Regenerative Endodontic Procedures with the Use of Chitosan Scaffolds in Immature Dog Teeth with Apical Periodontitis. *J. Endodont.* **2017**, *43*, 1279–1287. [CrossRef]
51. Neves, V.C.; Babb, R.; Chandrasekaran, D.; Sharpe, P.T. Promotion of natural tooth repair by small molecule GSK3 antagonists. *Sci. Rep.* **2017**, *7*, 39654. [CrossRef]
52. Mjor, I.A.; Tronstad, L. The healing of experimentally induced pulpitis. *Oral Surg. Oral Med. Oral Pathol.* **1974**, *38*, 115–121. [CrossRef] [PubMed]
53. Mjor, I.A. Pulp-dentin biology in restorative dentistry. Part 7: The exposed pulp. *Quintessence. Int.* **2002**, *33*, 113–135. [PubMed]
54. Galler, K.M. Clinical procedures for revitalization: Current knowledge and considerations. *Int. Endod. J.* **2016**, *49*, 926–936. [CrossRef] [PubMed]
55. Kim, S.G.; Zhou, J.; Solomon, C.; Zheng, Y.; Suzuki, T.; Chen, M.; Song, S.; Jiang, N.; Cho, S.; Mao, J.J. Effects of growth factors on dental stem/progenitor cells. *Dent. Clin. N. Am.* **2012**, *56*, 563–575. [CrossRef] [PubMed]
56. Washio, A.; Kitamura, C.; Morotomi, T.; Terashita, M.; Nishihara, T. Possible Involvement of Smad Signaling Pathways in Induction of Odontoblastic Properties in KN-3 Cells by Bone Morphogenetic Protein-2: A Growth Factor to Induce Dentin Regeneration. *Int. J. Dent.* **2012**, *2012*, 258469. [CrossRef]
57. Abd-Elmeguid, A.; Abdeldayem, M.; Kline, L.W.; Moqbel, R.; Vliagoftis, H.; Yu, D.C. Osteocalcin expression in pulp inflammation. *J. Endod.* **2013**, *39*, 865–872. [CrossRef]

58. Ishizaka, R.; Hayashi, Y.; Iohara, K.; Sugiyama, M.; Murakami, M.; Yamamoto, T.; Fukuta, O.; Nakashima, M. Stimulation of angiogenesis, neurogenesis and regeneration by side population cells from dental pulp. *Biomaterials* **2013**, *34*, 1888–1897. [CrossRef]
59. Iohara, K.; Zheng, L.; Wake, H.; Ito, M.; Nabekura, J.; Wakita, H.; Nakamura, H.; Into, T.; Matsushita, K.; Nakashima, M. A novel stem cell source for vasculogenesis in ischemia: Subfraction of side population cells from dental pulp. *Stem. Cells* **2008**, *26*, 2408–2418. [CrossRef]
60. Nakashima, M.; Iohara, K.; Sugiyama, M. Human dental pulp stem cells with highly angiogenic and neurogenic potential for possible use in pulp regeneration. *Cytokine Growth Factor Rev.* **2009**, *20*, 435–440. [CrossRef]
61. Huang, G.T.; Garcia-Godoy, F. Missing Concepts in De Novo Pulp Regeneration. *J. Dent. Res.* **2014**, *93*, 717–724. [CrossRef] [PubMed]
62. Meschi, N.; Hilkens, P.; Lambrichts, I.; Van den Eynde, K.; Mavridou, A.; Strijbos, O.; De Ketelaere, M.; Van Gorp, G.; Lambrechts, P. Regenerative endodontic procedure of an infected immature permanent human tooth: An immunohistological study. *Clin. Oral Investig.* **2016**, *20*, 807–814. [CrossRef] [PubMed]
63. Xu, W.N.; Jiang, S.; Chen, Q.Y.; Ye, Y.Y.; Chen, J.J.; Heng, B.C.; Jiang, Q.L.; Wu, B.L.; Ding, Z.H.; Zhang, C.F. Systemically Transplanted Bone Marrow-derived Cells Contribute to Dental Pulp Regeneration in a Chimeric Mouse Model. *J. Endodont.* **2016**, *42*, 263–268. [CrossRef] [PubMed]
64. Torabinejad, M.; Faras, H.; Corr, R.; Wright, K.R.; Shabahang, S. Histologic Examinations of Teeth Treated with 2 Scaffolds: A Pilot Animal Investigation. *J. Endodont.* **2014**, *40*, 515–520. [CrossRef]
65. Mullane, E.M.; Dong, Z.; Sedgley, C.M.; Hu, J.C.; Botero, T.M.; Holland, G.R.; Nor, J.E. Effects of VEGF and FGF2 on the revascularization of severed human dental pulps. *J. Dent. Res.* **2008**, *87*, 1144–1148. [CrossRef]
66. Choung, H.W.; Lee, J.H.; Lee, D.S.; Choung, P.H.; Park, J.C. The role of preameloblast-conditioned medium in dental pulp regeneration. *J. Mol. Histol.* **2013**, *44*, 715–721. [CrossRef]
67. Huang, G.T.; Yamaza, T.; Shea, L.D.; Djouad, F.; Kuhn, N.Z.; Tuan, R.S.; Shi, S. Stem/progenitor cell-mediated de novo regeneration of dental pulp with newly deposited continuous layer of dentin in an in vivo model. *Tissue Eng. Part A.* **2010**, *16*, 605–615. [CrossRef]
68. Nie, X.; Tian, W.; Zhang, Y.; Chen, X.; Dong, R.; Jiang, M.; Chen, F.; Jin, Y. Induction of transforming growth factor-beta 1 on dentine pulp cells in different culture patterns. *Cell Biol. Int.* **2006**, *30*, 295–300. [CrossRef]
69. Duncan, H.F.; Cooper, P.R.; Smith, A.J. Dissecting dentine-pulp injury and wound healing responses: Consequences for regenerative endodontics. *Int. Endod. J.* **2019**, *52*, 261–266. [CrossRef]
70. Laurent, P.; Camps, J.; About, I. Biodentine(TM) induces TGF-beta1 release from human pulp cells and early dental pulp mineralization. *Int. Endod. J.* **2012**, *45*, 439–448. [CrossRef]
71. Yu, C.; Abbott, P.V. An overview of the dental pulp: Its functions and responses to injury. *Aust. Dent. J.* **2007**, *52*, S4–S16. [CrossRef] [PubMed]
72. Chang, S.W.; Lee, S.Y.; Ann, H.J.; Kum, K.Y.; Kim, E.C. Effects of calcium silicate endodontic cements on biocompatibility and mineralization-inducing potentials in human dental pulp cells. *J. Endod.* **2014**, *40*, 1194–1200. [CrossRef] [PubMed]
73. Kim, J.; Song, Y.S.; Min, K.S.; Kim, S.H.; Koh, J.T.; Lee, B.N.; Chang, H.S.; Hwang, I.N.; Oh, W.M.; Hwang, Y.C. Evaluation of reparative dentin formation of ProRoot MTA, Biodentine and BioAggregate using micro-CT and immunohistochemistry. *Restor. Dent. Endod.* **2016**, *41*, 29–36. [CrossRef] [PubMed]
74. Rajasekharan, S.; Martens, L.C.; Vandenbulcke, J.; Jacquet, W.; Bottenberg, P.; Cauwels, R.G. Efficacy of three different pulpotomy agents in primary molars: A randomized control trial. *Int. Endod. J.* **2017**, *50*, 215–228. [CrossRef]
75. Cardoso, M.; Dos Anjos Pires, M.; Correlo, V.; Reis, R.; Paulo, M.; Viegas, C. Biodentine for Furcation Perforation Repair: An Animal Study with Histological, Radiographic and Micro-Computed Tomographic Assessment. *Iran. Endod. J.* **2018**, *13*, 323–330. [CrossRef]
76. Noetzel, J.; Ozer, K.; Reisshauer, B.H.; Anil, A.; Rossler, R.; Neumann, K.; Kielbassa, A.M. Tissue responses to an experimental calcium phosphate cement and mineral trioxide aggregate as materials for furcation perforation repair: A histological study in dogs. *Clin. Oral Investig.* **2006**, *10*, 77–83. [CrossRef]
77. Aktemur Turker, S.; Uzunoglu, E.; Deniz Sungur, D.; Tek, V. Fracture Resistance of Teeth with Simulated Perforating Internal Resorption Cavities Repaired with Different Calcium Silicate-based Cements and Backfilling Materials. *J. Endod.* **2018**, *44*, 860–863. [CrossRef]
78. Rajasekharan, S.; Martens, L.C.; Cauwels, R.; Anthonappa, R.P.; Verbeeck, R.M.H. Biodentine material characteristics and clinical applications: A 3 year literature review and update. *Eur. Arch. Paediatr. Dent.* **2018**, *19*, 1–22. [CrossRef]
79. Nowicka, A.; Wilk, G.; Lipski, M.; Kolecki, J.; Buczkowska-Radlinska, J. Tomographic Evaluation of Reparative Dentin Formation after Direct Pulp Capping with $Ca(OH)_2$, MTA, Biodentine, and Dentin Bonding System in Human Teeth. *J. Endod.* **2015**, *41*, 1234–1240. [CrossRef]
80. De Rossi, A.; Silva, L.A.; Gaton-Hernandez, P.; Sousa-Neto, M.D.; Nelson-Filho, P.; Silva, R.A.; de Queiroz, A.M. Comparison of pulpal responses to pulpotomy and pulp capping with biodentine and mineral trioxide aggregate in dogs. *J. Endod.* **2014**, *40*, 1362–1369. [CrossRef]
81. Shayegan, A.; Jurysta, C.; Atash, R.; Petein, M.; Abbeele, A.V. Biodentine used as a pulp-capping agent in primary pig teeth. *Pediatr. Dent.* **2012**, *34*, e202–e208. [PubMed]

82. Feng, X.; Feng, G.; Xing, J.; Shen, B.; Li, L.; Tan, W.; Xu, Y.; Liu, S.; Liu, H.; Jiang, J.; et al. TNF-alpha triggers osteogenic differentiation of human dental pulp stem cells via the NF-kappaB signalling pathway. *Cell Biol. Int.* **2013**, *37*, 1267–1275. [CrossRef] [PubMed]
83. Zhao, Y.; Wang, C.L.; Li, R.M.; Hui, T.Q.; Su, Y.Y.; Yuan, Q.; Zhou, X.D.; Ye, L. Wnt5a promotes inflammatory responses via nuclear factor kappaB (NF-kappaB) and mitogen-activated protein kinase (MAPK) pathways in human dental pulp cells. *J. Biol. Chem.* **2014**, *289*, 21028–21039. [CrossRef] [PubMed]

Journal of
Functional Biomaterials

MDPI

Article

Microleakage Evaluation of Temporary Restorations Used in Endodontic Treatment—An Ex Vivo Study

Siri Paulo [1,2,3,*], Ana Margarida Abrantes [2,3,4,5], Mariana Xavier [1], Ana Filipa Brito [2,5], Ricardo Teixo [2,3,5], Ana Sofia Coelho [2,3,4,6], Anabela Paula [2,3,4,6], Eunice Carrilho [2,3,4,6], Maria Filomena Botelho [2,3,4,5], Carlos Miguel Marto [2,3,4,5,6,7] and Manuel Marques Ferreira [1,2,3,4]

1 Institute of Endodontics, Faculty of Medicine, University of Coimbra, 3000-548 Coimbra, Portugal
2 Coimbra Institute for Clinical and Biomedical Research (iCBR), Area of Environment Genetics and Oncobiology (CIMAGO), Faculty of Medicine, University of Coimbra, 3000-548 Coimbra, Portugal
3 Center for Innovative Biomedicine and Biotechnology (CIBB), University of Coimbra, 3000-548 Coimbra, Portugal
4 Clinical Academic Center of Coimbra (CACC), 3004-561 Coimbra, Portugal
5 Institute of Biophysics, Faculty of Medicine, University of Coimbra, 3000-548 Coimbra, Portugal
6 Institute of Integrated Clinical Practice, Faculty of Medicine, University of Coimbra, 3000-075 Coimbra, Portugal
7 Institute of Experimental Pathology, Faculty of Medicine, University of Coimbra, 3000-548 Coimbra, Portugal
* Correspondence: sirivpaulo@gmail.com

Abstract: (1) Background: Coronal microleakage can lead to endodontic treatment failure. This study aimed to compare the sealing ability of different temporary restorative materials used during endodontic treatment. (2) Methods: Eighty sheep incisors were collected, uniformized in length, and access cavities were performed, except for in the negative control group, where the teeth were left intact. The teeth were divided into six different groups. In the positive control group, the access cavity was made and left empty. In the experimental groups, access cavities were restored with three different temporary materials (IRM®, Ketac™ Silver, and Cavit™) and with a definitive restorative material (Filtek Supreme™). The teeth were submitted to thermocycling, and two and four weeks later, they were infiltrated with $^{99m}TcNaO_4$, and nuclear medicine imaging was performed. (3) Results: Filtek Supreme™ obtained the lowest infiltration values. Regarding the temporary materials, at two weeks, Ketac™ Silver presented the lowest infiltration, followed by IRM®, whereas Cavit™ presented the highest infiltration. At four weeks, Ketac™ Silver remained with the lowest values, whereas Cavit™ decreased the infiltration, comparable to IRM®. (4) Conclusion: Regarding temporary materials, Ketac™ Silver had the lowest infiltration at 2 and 4 weeks, whereas the highest infiltration was found in the Cavit™ group at two weeks and in the IRM® group at 4 weeks.

Keywords: endodontic treatment; sealing ability; microleakage; temporary material; nuclear medicine

1. Introduction

Endodontic treatment is based on chemical and mechanical debridement, root canal filling, and later definitive crown restoration, to eliminate bacteria and prevent reinfection [1–4]. Before the treatment, it is fundamental to remove dental caries and infiltrated old restorations, or perform pre-endodontic restorations, to avoid microleakage, which is defined as the diffusion of saliva, microorganisms and their products, ions, and molecules to the canals, which can lead to treatment failure [3,5–8]. Several reports identified bacteria strains from the oral cavity, such as *Prevotella intermedia, Porphyromonas gingivalis, Enterococcus faecalis, Streptococcus viridans, Staphylococcus* sp., and *Enterococcus faecalis*, as responsible for persistent infections and endodontic failures after endodontic treatment [1,7,9].

To avoid reinfection during or after the treatment and consequently improve prognosis, a proper provisional restoration must be performed during the endodontic treatment [1].

Even after the root canal filling, until the final restoration is performed, temporary fillings are still essential since the obturated root canal system, exposed to saliva, is also susceptible to microleakage [10,11]. An infiltration of 79–85% of the obturated root canal system in a 3- to 56-day interval [12], and a microleakage of *S. epidermidis* in 19 days and of *P. vulgaris* in 42 days in teeth without an efficient restoration [10,11] was reported, supporting the importance of temporary restorations for treatment success.

The ideal temporary material should avoid contact between the root canal system and the oral environment and be resilient to abrasion and compression. It should display low porosity, dimensional stability, good sealing, and reasonable aesthetics, and have the capacity to prevent the canal system from becoming contaminated by saliva, fluids, and microorganisms [3,5,6,13]. In addition, to be an effective barrier material, a minimum of 4 mm thickness is necessary [1,5,14–16]. Nowadays, the temporary materials most used in clinical practice are Cavit™ (3M ESPE, Seefeld, Germany), a calcium sulfate-based cement; Ketac Silver™ (3M ESPE, Seefeld, Germany), a glass ionomer cement with silver particles; and IRM®, a reinforced zinc-oxide-eugenol cement (Dentsply, Milford, DE, USA). The choice of the material to be used depends on the clinical needs of each case, such as duration of use, dimensional stability, abrasion resistance, stabilization of intracanal medication, and adaptation to more complex access cavity formats [3,17]. Several studies, mostly in vitro, have evaluated the seal capacity of these materials, and there is evidence that microleakage occurs to different degrees for most temporary materials available, and none of them can entirely prevent microleakage from the 1st to the 14th day [1–4,18].

Several studies have been performed on this topic, but the obtained data presents contradictory results [1,3,4,19–23]. This may be due to different testing materials, evaluated time points, and experimental protocols [4,14,20–25]. However, the main reason is the use of different methodologies to appraise microleakage, namely dyes, radioisotopes, bacteria, or their sub-products [1,26,27]. Most studies use dyes such as methylene blue, but more sensitive methods, such as nuclear medicine, can provide more accurate results [26,27].

Considering its influence on the treatment's success, evaluating which materials can successfully prevent microleakage is fundamental. Therefore, the three most used temporary restoration materials and a definitive restoration material were compared in the present study. The null hypothesis was that there were no differences in microleakage between the three temporary restorative materials—Cavit™, Ketac Silver™, and IRM® at 2 and 4 weeks. A second null hypothesis was that there were no differences in microleakage between the temporary materials and the definitive restorative material—composite resin Filtek Supreme™.

2. Materials and Methods

2.1. Sample Preparation

For this study, sheep teeth were used. The teeth were obtained post mortem at a food sector abattoir. The animals were handled and euthanized according to the Portuguese (DL 98/96, Art. 1°) and European Legislation concerning animal welfare (EFSA, AHAW/04-027). Eighty incisor teeth from two-year-old sheep were collected and cleaned by removing soft tissues and other residues. The teeth were then disinfected in azide chloride solution (0.01 g/mL) for three days [27].

To obtain uniformized samples of 16 mm in length, the teeth were sectioned 2 mm above the cementoenamel junction, except in the negative control group. After this, [4,6,20,27] a spherical diamond turbine drill with a 2.1 mm diameter at high speed and constant irrigation was used to create access cavities of 2.1 mm (length) × 2.1 mm (width) × 4–4.5 mm (height) [18], measured with a periodontal probe, to allow the minimum thickness of 4 mm for the restoration material (Figure 1a) [1,5,14–16]. The remaining pulp tissue was removed using an endodontic K30 file with sodium hypochlorite (2.5%) irrigation. Next, a glide path with manual files K10 and K15 was achieved, followed by canal instrumentation using the ProTaper NEXT™ (PTN) system (Dentsply Sirona), triggered by X-SMART™ (Dentsply Sirona) with a constant rotation of 300 RPM and a torque of 3. A sequence of the PTN™

files PG, X1, X2, X3, and X4 or X5 was used, depending on the diameter of the different canals (Figure 1a). Next, irrigation with sodium hypochlorite (2.5%) and permeabilization with a manual K10 file were performed between each file. Following this, the canals were irrigated with EDTA (15%) and then with 3 mL of 0.9% saline solution for 3 min to neutralize EDTA. Finally, all the canals were dried with paper points.

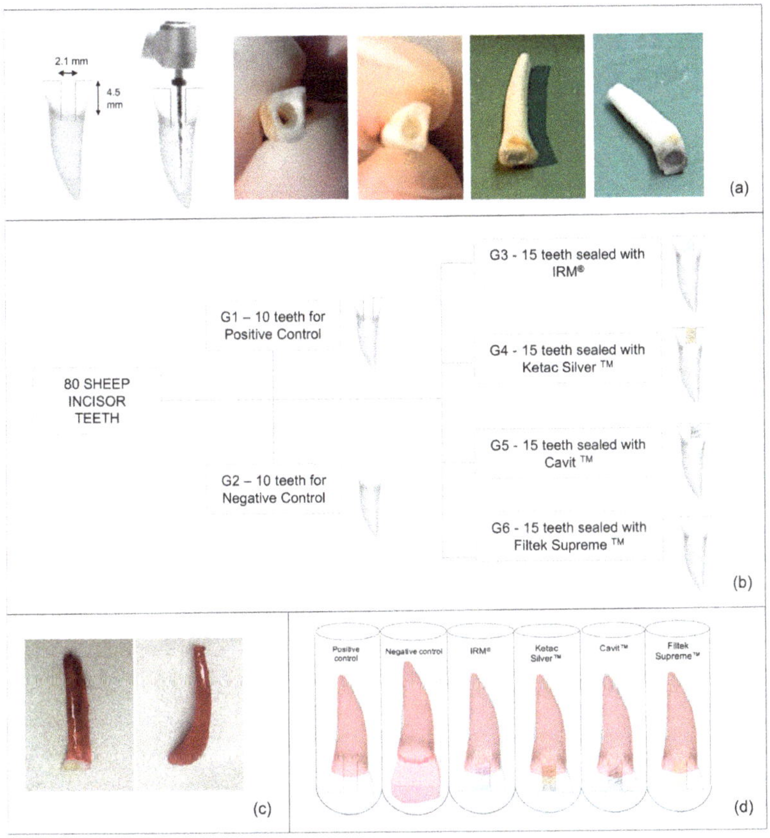

Figure 1. Study flowchart. (**a**). Preparation of uniformized samples—teeth were sectioned 2 mm above the cementoenamel junction; an access cavity was made with a round bur; teeth were instrumented and had Teflon condensed at the bottom of the cavities, leaving a 4 mm height to restoration placement. (**b**). Groups distribution. (**c**). Preparation for nuclear medicine analysis—tooth impermeabilized with two layers of nail polish, in all root surface besides in the last 1 mm of the access cavities (except in the negative group in which all the tooth surface was covered). (**d**). Teeth in test tubes, immersed in $^{99m}TcNaO_4$.

Finally, the teeth were randomly divided into two groups of 10 for the positive and negative control groups and four experimental groups of 15 teeth each (Figure 1b). Then, Teflon was placed in the access cavities and condensed at the cavities' bottom, at the canal's entrance, to support the temporary material [28].

2.2. Temporary Restoration Material Fillings

Except for Group 1 (teeth with empty access cavity—positive control) and for Group 2 (intact teeth—negative control), the remaining access cavities were filled with different temporary restorative materials (Table 1).

Table 1. Study groups.

Group	Material	Manufacturer	Composition *	Lot Number
1	No material placement	-	-	-
2	Intact teeth	-	-	-
3	IRM®	Dentsply Sirona Inc. Milford, DE, USA	powder: zinc oxide, poly-methyl methacrylate (PMMA) powder, pigment liquid: eugenol, acetic acid	powder: 1910001036 liquid: 2001000680
4	Ketac Silver™	3M ESPE, Seefeld, Germany	powder: silver, oxide glass chemicals (non-fibrous), titanium dioxide, copper liquid: water, copolymer of acrylic acid—maleic acid, tartaric acid	7964610
5	Cavit™	3M ESPE, Seefeld, Germany	zinc oxide; sulfuric acid, calcium salt, hydrate; ethylene bis(oxyethylene) diacetate; zinc sulfate; poly(vinyl acetate)	7121289
6	Filtek Supreme™ (+Scotchbond™ Universal)	3M ESPE, Seefeld, Germany	bis-GMA, UDMA, TEGDMA, and bis-EMA resins (MDP phosphate monomer, dimethacrylate resins, HEMA, Vitebond™ copolymer, filler, ethanol, water, initiators, silane)	NC45009 (7676507)

* Information provided by the manufacturer.

The materials were prepared and used following the manufacturer's instructions. For IRM®, a 1:1 mixture was prepared until a stable consistency was attained. The polymerization occurred after a few seconds. Ketac Silver™ was activated with vibration for 10 s. The material was then applied with a Ketac Aplicap™ Applicator (3M ESPE, Seefeld, Germany). This is a self-cure material with a short working time, and its setting starts within seconds. Cavit™ is a pre-mixed material that does not require any preparation. It was taken directly from the container, applied to the tooth, and left to set for 4 h. Before the restoration with Filtek Supreme™, a selective enamel etching with orthophosphoric acid (37%) was performed, and Scotchbond™ Universal dental adhesive (3M ESPE, Seefeld, Germany) was applied, according to the manufacturer's instructions. The composite was inserted in the cavities in 2 mm layers. The adhesive system and each layer of the composite resin was photopolymerized.

The same experienced operator performed all the temporary restorations to avoid technical bias. Upon the crown sealing, all teeth were placed in 0.9% saline solution at room temperature to mimic the clinical environment [14,23,25,27].

2.3. Thermocycling

The teeth were subjected to thermic stress to simulate the aging of the restoration materials that occurs clinically. For this step, teeth were alternately placed in baths at 5 ± 5 °C and 55 ± 5 °C for 30 s periods in each bath, totaling 500 cycles. These temperatures, intervals, and the number of cycles were chosen according to the ISO/TS 11405: 2015 recommendations. The 30 s interval mimics the latency time of the oral environment to recover its normal temperature after exposure to hot or cold food or drinks [29–31]. The thermocycling set-up was chosen considering 10,000 cycles/year occurs. After thermocycling, the samples were kept in saline until nuclear medicine analysis.

2.4. Nuclear Medicine

Two and four weeks after thermocycling, all samples were dried, sealed apically with cyan acrylate [3] and impermeabilized with 2 layers of nail polish (Catrice Cosmetics) in all root surface besides in the last 1 mm of the access cavities, except in the negative group in which all the tooth surface was covered [3,15] as depicted in Figure 1c.

Impermeabilized teeth were placed in a sodium pertechnetate solution, $^{99m}TcNaO_4$ (8 mCi/mL), immersing only the non-impermeabilized segment for 3 h (Figure 1d) [27]. Teeth were then washed in tap water at constant flow for 30 s each. After this, the teeth were dried with absorbent paper, and the nail polish was removed with a scalpel.

Next, for each sample, a 512×512 pixel image was acquired in a gamma chamber (Millennium, New York, NY, USA) for 2 min. In each image, a region of interest (ROI) with the same pixel size was applied to obtain the total, maximum, and average values of micro infiltration. After the two-week analysis, the samples were again stored in the saline solution until the four-week evaluation [14,23,25,27]. At four weeks, the described nuclear medicine analysis was repeated.

2.5. Statistical Analysis

The sample size of 15 specimens in experimental groups and 10 specimens in control groups was chosen in accordance with previously published similar studies [4,6,21,32]. The sample size was validated by a post hoc analysis of sample size calculation, which retrieved a confidence interval of 99.15%, and a size effect of 2.4, using G* Power 3.1 [33].

Statistical analysis was performed using IBM SPSS Statistics 28 software (IBM Corporation, Armonk, NY, USA) and data were presented as mean ± standard deviation (SD). The Shapiro–Wilk test was used to evaluate the normality of each population distribution, and the Levene test was used to assess the variance homogeneity. Parametric tests (ANOVA) were employed when normal distribution was observed, and non-parametric tests (Kruskal–Wallis) were used for non-normal distributions to assess multiple comparisons of the different temporary materials studied at the same time of assessment, with post hoc correction by Tukey's test. To compare each of the different temporary materials at different evaluation times, the parametric Student's *t*-test and the non-parametric Wilcoxon test were used according to sample normality. A family-wise 95% confidence level ($p < 0.05$) was applied to all statistical analyses.

3. Results

At 2 and 4 weeks, the positive group, i.e., teeth with cavity access performed but without restoration, displayed the highest counts per minute (cpm), indicating the highest microleakage across the different groups (Figures 2–4, Table 2).

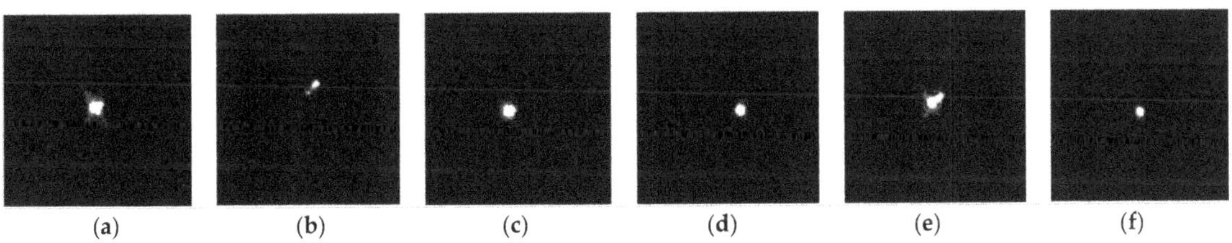

Figure 2. Representative scintigraphy images of positive control (**a**), negative control (**b**), IRM® (**c**), Ketac™ (**d**), Cavit™ (**e**), and Filtek™ (**f**) groups with uptake of $^{99m}TcNaO_4$.

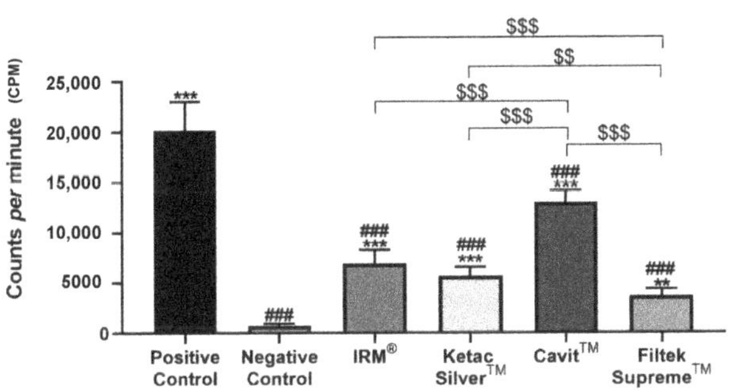

Figure 3. Total counts per minute (cpm) were obtained after infiltration with $^{99m}TcNaO_4$, 2 weeks after restoration. ### means $p < 0.001$ (relative to the positive control); ** means $p < 0.01$ and *** means $p < 0.001$ (relative to the negative control); $$ means $p < 0.01$ and $$$ means $p < 0.001$ (comparisons between materials). Data are presented as mean ± standard deviation (SD).

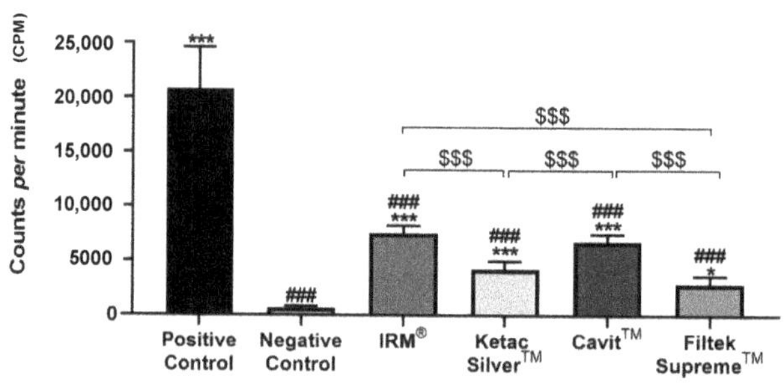

Figure 4. Total counts per minute (cpm) were obtained after infiltration with $^{99m}TcNaO_4$, 4 weeks after restoration. ### means $p < 0.001$ (relative to the positive control); * means $p < 0.05$, and *** means $p < 0.001$ (relative to the negative control); $$$ means $p < 0.001$ (comparisons between materials). Data are presented as mean ± SD.

Table 2. Counts per minute of control and experimental groups after 2 and 4 weeks. Data are presented as mean ± standard deviation. When comparing 4 weeks vs. 2 weeks within each group, ** means $p < 0.01$, and *** means $p < 0.001$.

	G1—Positive Control	G2—Negative Control	G3—IRM®	G4—Ketac™ Silver	G5—Cavit™	G6—Filtek Supreme™
2 weeks	20,181.0 ± 2858.9	697.8 ± 222.0	6871.8 ± 1360.2	5615.4 ± 912.3	12,952.9 ± 1199.6	3620.9 ± 702.7
4 weeks	20,769.1 ± 3865.5	669.3 ± 143.3	7561.1 ± 677.2	4236.9 ± 755.6	6807.9 ± 660.2	2927.6 ± 735.1
p-value 4 vs. 2 weeks	0.575	0.845	0.114	0.002 **	<0.001 ***	0.059

Conversely, the negative control group, composed of intact teeth without restoration, displayed the lowest levels of infiltration at the two time points (Figures 3 and 4, Table 2). At both times, all tested materials presented a significantly lower infiltration than the positive control group and a higher infiltration relative to the negative control group (Figures 3 and 4, Table 2).

At 2 weeks after restoration, Cavit™ displayed the highest values of $^{99m}TcNaO_4$ infiltration, significantly higher than the other three tested materials, followed by IRM®. (Table 2, Figure 3).

At 4 weeks, Cavit™ and IRM® showed similar infiltration, higher than that for Ketac™ Silver, which showed the lowest infiltration observed in the teeth with temporary restoration, identical to that observed at 2 weeks (Figure 4).

Comparing the infiltration between 2 and 4 weeks within each group, a significant reduction in counts per minute was found from 2 to 4 weeks in Ketac Silver™ and with Cavit™ groups, but not in Filtek Supreme™ or with IRM®. No changes in infiltration from 2 to 4 weeks were observed in both positive and negative control groups (Table 2).

4. Discussion

The present study aimed to compare the sealing ability of three widely used temporary restorative materials. Temporary restorations are fundamental during and after endodontic treatment to prevent microleakage and ensure treatment success. Conversely, the lack of satisfactory temporary restorations can lead to bacteria infiltration, flare-up reactions, antibiotics administration, and, ultimately, a poor prognosis [1,17].

Regarding the experimental model chosen, using ex vivo models in dentistry is widespread and helpful [34]. Therefore, sheep teeth were selected since they show anatomic and histological similarities with human teeth and are commonly used in endodontic research [34–36]. In addition, they allow the necessary sample to be obtained in a reasonable time, which is a disadvantage of using human teeth, especially specific teeth such as incisors.

During the experimental protocol, thermocycling was chosen to simulate the oral conditions, namely hot and cold food exposure [30]. In the oral cavity, temperature changes affect the marginal seal of restoration materials because the linear coefficients of thermal expansion of the materials and dental tissues are different [22,37]. This way, the temperature changes altered the linear coefficient of thermal expansion of both the tooth and the restorative material, mimicking oral conditions. In addition, the artificial aging induced by thermocycling includes an initial phase with hot water, accelerating the hydrolysis of unprotected collagen [38]. Consequently, the high coefficient of thermal expansion/contraction of the material generates stress in the tooth-restoration interface and creates spaces that favor the pathway to fluids and microorganisms [31]. Additionally, the teeth were stored in a saline solution until being evaluated. As suggested by ISO/TS 11405: 2015, storage in an aqueous medium allows the differentiating of materials that resist wet environments from those that do not. Since all materials are used in the oral cavity, which is a moist environment, the ability to resist wet environments is essential [21]. In addition, the two time points chosen simulate a period between appointments of 15 days, similar to clinical conditions.

The obtained results show all tested materials presented microleakage, although to different degrees. The first null hypothesis was rejected since, at 2 weeks, Cavit™ presented a significantly higher microleakage when compared with IRM® and Ketac™ Silver. At 4 weeks, Ketac™ Silver presented significantly lower microleakage compared with IRM® and with Cavit™. The second null hypothesis was also rejected since composite resin Filtek Supreme™ presented significantly lower microleakage at 2 weeks compared with the three temporary restorative materials, and at 4 weeks, when compared with IRM® and with Cavit™.

Overall, Cavit™ presented significantly high levels of infiltration at 2 weeks, which decreased abruptly at 4 weeks, whereas IRM® microinfiltration remained stable from 2 to 4 weeks. Ketac™ Silver presented significantly lower infiltration at 2 and 4 weeks.

Cavit™ is a pre-mixed auto-polymerized material that contains zinc oxide and synthetic resins without eugenol. Cavit™ is prefabricated in three different forms: Cavit™ (pink), Cavit™-W (white), and Cavit™-G (grey). Cavit™ and Cavit™-W have different concentrations of zinc sulfate and zinc oxide, resulting in a hardness increase for Cavit™ and an adhesion increase for Cavit™-W [2]. In this study, Cavit™ was used, which has an indication to be used during endodontic treatment. Cavit™ is endowed with favorable properties for crown sealing, such as hygroscopic expansion due to water absorption [2]. However, it is also characterized by a low mechanical resistance and slow setting [2,15]. Other studies evaluating this material present contradictory results, probably related to experimental set-up and evaluation periods [3,4,22,23,39,40]. In the present study, Cavit™ presented an abrupt decrease in infiltration values from 2 to 4 weeks. The slow setting may justify the high infiltration at 2 weeks. At 4 weeks, the hygroscopic properties and the elevated coefficient of linear expansion resulting from water absorption (approximately double that of IRM™) seem to contribute to the sealing improvement and infiltration decrease. Nevertheless, microleakage was superior to other materials at both time points, and thus it is not an appropriate material for longer periods. If a longer time period were to be evaluated, infiltration would be expected to increase again since the existence of a high number of pores facilitates water absorption and adhesion degradation over time [22].

IRM® is an auto-polymerized material that contains zinc oxide and eugenol. Although it is characterized by more difficult manipulation [15], it showed intermediate levels of infiltration. As for Cavit™, several studies present non-consensual results for IRM® infiltration. Similar to our results, some show IRM®'s sealing ability to be superior to that of Cavit™ at shorter evaluation times, whereas others present IRM®'s sealing ability to be inferior to that of Cavit™ [4,18]. Again, different experimental setups and evaluating methodologies can explain the contradictory results. In the present study, IRM®'s infiltration values slightly increased from 2 to 4 weeks, without statistical difference. That can be related to the continuous release of eugenol, which is hydro-soluble and favors the detachment of particles from IRM®, increasing porosity [41]. In addition, the higher level of infiltration, when compared with that of Ketac™, can be related to a variation of volume arising from polymerization contraction or to the non-homogeneous handmade mixture subjected to operator variability [23,42]. Nevertheless, this material demonstrated stability over the evaluation periods.

Ketac™ is a metal-reinforced glass ionomer material with fine particles of silver fused with the glass. It has an ion-leachable alumino-silicate glass powder that combines with poly-alkeonic acid liquid. When mixed, it auto-polymerizes, releasing aluminum and calcium ions that form salt bridges and a gel matrix that adheres to mineralized tissues [43,44]. Adhesion to the teeth's surface is related to ionic forces associated with chelation with calcium [23]. Due to this, Ketac™ is characterized by good adhesion and resistance to fracture [45]. The contraction is initially a slow cross-linked reaction that enables stress relief and maintains a homogeneous and consistent interface between the tooth and the material, leading to reliable adhesion [20]. Despite its high sensibility to manipulation and contraction during polymerization, which sometimes impairs good results, in the present study, this material showed lower levels of infiltration in comparison with Cavit™ and IRM®, most likely due to the referred chemical adhesion capacity.

As expected, the resin composite was the material that best prevented microleakage. In this study, an enamel selective etching procedure and a universal dental adhesive containing 10-MDP (Scotchbond™ Universal) were used. The use of functional monomers (such as 10-MDP) in dental adhesives promotes a chemical interaction with both dentin and enamel, improving bond strength and the restoration's quality and longevity [46–48]. The reliable characteristic adhesion of a composite restoration to the tooth structure justified the results. It showed that although temporary materials achieved reasonable results, none of them could present low microleakage results as the ones of definitive restorative materials. This supports the recommendation to perform crown restorations with composites around the access cavity, if necessary, to reduce fluid infiltration at minimum during endodontic treatment [3].

As previously stated, published literature on the topic of temporary restorations presents discrepant results. For instance, several studies reported Cavit™ presenting lower infiltration values than IRM® and Ketac™ [15,19,20,40]. Different methodologies may underlie the discrepancies reported in the different studies [14,16,19,49]. It can be explained by the use of different evaluation methods and times and different experimental set-ups to simulate oral conditions, namely thermocycling or/and cyclic loading, the number of cycles in thermocycling, and the materials tested [14,16,19,49].

The infiltration evaluation can be performed using several methodologies, such as methylene blue dye, radioactive isotopes, bacteria, or fluid filtration. However, the use of dyes, mainly methylene blue, is the most frequently used because it is a sensitive indicator of infiltration as it has a small molecule size, similar to the size of microorganisms [14,19]. It has the advantage of not being absorbed by the hydroxyapatite crystal of dentin. Nevertheless, its accuracy depends on how much air is entrapped in the restoration [22]. In addition, this technique has the disadvantage of sectioning the tooth to assess the amount of microleakage, implying a meticulous standardization of the sectioned portions [26]. Bacterial studies are also widely used. However, several different microorganisms are responsible for pulpal infection, and with this method, only a few are evaluated [20]. In the present study, the microleakage was assessed using the infiltration of radioisotope $^{99m}TcNaO_4$, a methodology that allows the monitorization and quantification of microleakage at different time points [26,27]. Technetium is a radionuclide with a smaller molecular size comparable to that of the microorganisms present in saliva. It presents selectivity, traveling through the tooth by capillarity and depositing in the "free" areas [26]. It has the advantage over the use of dyes of not requiring the sectioning of the sample, thus allowing the evaluation of the same sample at different time points. This is possible because $^{99m}TcNaO_4$ will decay to $^{99}TcNaO_4$ which is a stable molecule. In this transition, only energy is dissipated and the molecule does not change. In addition, it presents higher sensitivity compared with the other methods, considering the very small concentrations needed, thus providing more accurate quantitative measures [50]. In fact, the sensitivity is so high that it is only necessary for one atom to be infiltrated for it be detected using a gamma camera. Its main disadvantage is the necessity of specific equipment and radiopharmaceutical availability, which limits its use. Importantly, the results obtained for the positive and negative groups validate the experimental model.

Although the experimental set-up tried to mimic clinical conditions, not all oral variables were reproduced, such as masticatory and occlusal load, which is a limitation. Other analyses, such as push-out tests, could also provide a more comprehensive evaluation. In addition, although the oral cavity has saliva permanently, the restorations could not be submerged entirely but instead were kept in a moist environment, which differs from the protocol used. Other evaluation times could allow more relevant information to be obtained—for instance, shorter times such as 24 h and 7 days. In addition, the increase in thermocycling cycles could better mimic situations where the temporary restorations stay longer in the oral cavity. Finally, in further studies, the use of micro-computed tomography, with superior resolution, will allow a tri-dimensional image of gaps at the tooth/restorative material interface to identify the areas where the restoration/tooth interface fails [51].

Nevertheless, the results obtained in this study are significant and reinforce that temporary restorations should only be used in specific clinical situations and for a short time since their efficacy to prevent microleakage is significantly lower than that of definitive materials. After completing the endodontic treatment, a definitive restoration should be placed as soon as possible.

5. Conclusions

Considering the limitations of this ex vivo study, it can be concluded that the three temporary materials present microleakage at 2 and 4 weeks. The experimental hypothesis was rejected since the temporary materials presented different microleakage. Ketac™ Silver presented the lowest infiltration at 2 and 4 weeks, whereas Cavit™ presented a higher infiltration at 2 weeks, diminishing sharply at 4 weeks. At this time, IRM® and Cavit™ prsented similar results. In addition, the second experimental hypothesis was rejected because the definitive material, resin composite, presented significantly lower microleakage than all tested temporary materials.

Author Contributions: Conceptualization, S.P., A.M.A., M.X. and M.M.F.; Data curation, A.M.A., M.X., A.F.B. and R.T.; Formal analysis, A.F.B., R.T., A.S.C., A.P. and C.M.M.; Funding acquisition, E.C., M.F.B. and M.M.F.; Methodology, S.P., A.M.A., A.P. and M.M.F.; Supervision, E.C. and M.F.B.; Writing—original draft, S.P., M.X. and C.M.M.; Writing—review & editing, A.M.A., A.S.C. and C.M.M. All authors have read and agreed to the published version of the manuscript.

Funding: This research was funded by Foundation for Science and Technology (FCT), Portugal. CIBB is funded by FCT Strategic Projects Strategic Projects UIDB/04539/2020 and UIDP/04539/2020 (CIBB) and UIDB/00102/2020.

Data Availability Statement: Data are contained within the article.

Conflicts of Interest: The authors declare no conflict of interest.

References

1. Shanmugam, S.; PradeepKumar, A.R.; Abbott, P.V.; Periasamy, R.; Velayutham, G.; Krishnamoorthy, S.; Mahalakshmi, K. Coronal Bacterial Penetration after 7 Days in Class II Endodontic Access Cavities Restored with Two Temporary Restorations: A Randomised Clinical Trial. *Aust. Endod. J.* **2020**, *46*, 358–364. [CrossRef]
2. Djouiai, B.; Wolf, T.G. Tooth and Temporary Filling Material Fractures Caused by Cavit, Cavit W and Coltosol F: An In Vitro Study. *BMC Oral Health* **2021**, *21*, 74. [CrossRef]
3. Nagpal, A.; Srivastava, P.K.; Setya, G.; Chaudhary, A.; Dhanker, K. Assessment of Coronal Leakage of Temporary Restorations in Root Canal-Treated Teeth: An In Vitro Study. *J. Contemp. Dent. Pract.* **2017**, *18*, 126–130. [CrossRef]

4. Babu, N.S.V.; Bhanushali, P.V.; Bhanushali, N.V.; Patel, P. Comparative Analysis of Microleakage of Temporary Filling Materials Used for Multivisit Endodontic Treatment Sessions in Primary Teeth: An In Vitro Study. *Eur. Arch. Paediatr. Dent.* **2019**, *20*, 565–570. [CrossRef]
5. Rechenberg, D.-K.; Schriber, M.; Attin, T. Bacterial Leakage through Temporary Fillings in Core Buildup Composite Material—An In Vitro Study. *J. Adhes. Dent.* **2012**, *14*, 371–376. [CrossRef]
6. Bilgrami, A.; Alam, M.K.; Qazi, F.U.R.; Maqsood, A.; Basha, S.; Ahmed, N.; Syed, K.A.; Mustafa, M.; Shrivastava, D.; Nagarajappa, A.K.; et al. An In-Vitro Evaluation of Microleakage in Resin-Based Restorative Materials at Different Time Intervals. *Polymers* **2022**, *14*, 466. [CrossRef]
7. Siqueira, J.F.; Rôças, I.N. Present Status and Future Directions: Microbiology of Endodontic Infections. *Int. Endod. J.* **2022**, *55*, 512–530. [CrossRef]
8. Muliyar, S.; Shameem, K.A.; Thankachan, R.P.; Francis, P.G.; Jayapalan, C.S.; Hafiz, K.A.A. Microleakage in Endodontics. *J. Int. Oral Health* **2014**, *6*, 99–104.
9. Alghamdi, F.; Shakir, M. The Influence of Enterococcus Faecalis as a Dental Root Canal Pathogen on Endodontic Treatment: A Systematic Review. *Cureus* **2020**, *12*, e7257. [CrossRef]
10. Naoum, H.J.; Chandler, N.P. Temporization for Endodontics. *Int. Endod. J.* **2002**, *35*, 964–978. [CrossRef]
11. Sritharan, A. Discuss that the Coronal Seal Is More Important than the Apical Seal for Endodontic Success. *Aust. Endod. J.* **2002**, *28*, 112–115. [CrossRef]
12. Swanson, K.; Madison, S. An Evaluation of Coronal Microleakage in Endodontically Treated Teeth. Part I. Time Periods. *J. Endod.* **1987**, *13*, 56–59. [CrossRef]
13. Tennert, C.; Fischer, G.F.; Vach, K.; Woelber, J.P.; Hellwig, E.; Polydorou, O. A Temporary Filling Material during Endodontic Treatment May Cause Tooth Fractures in Two-Surface Class II Cavities In Vitro. *Clin. Oral Investig.* **2016**, *20*, 615–620. [CrossRef]
14. Chailertvanitkul, P.; Abbott, P.V.; Riley, T.V.; Sooksuntisakoonchai, N. Bacterial and Dye Penetration through Interim Restorations Used during Endodontic Treatment of Molar Teeth. *J. Endod.* **2009**, *35*, 1017–1022. [CrossRef]
15. Çiftçi, A.; Vardarlı, D.A.; Sönmez, I.Ş. Coronal Microleakage of Four Endodontic Temporary Restorative Materials: An In Vitro Study. *Oral Surg. Oral Med. Oral Pathol. Oral Radiol. Endodontol.* **2009**, *108*, e67–e70. [CrossRef]
16. Adnan, S.; Khan, F.R. Comparison of Micro-Leakage around Temporary Restorative Materials Placed in Complex Endodontic Access Cavities: An In-Vitro Study. *J. Coll. Physicians Surg. Pak.* **2016**, *26*, 182–186.
17. Sivakumar, J.; Suresh Kumar, B.; Shyamala, P. Role of Provisional Restorations in Endodontic Therapy. *J. Pharm. Bioallied Sci.* **2013**, *5*, 120. [CrossRef]
18. Markose, A.; Krishnan, R.; Ramesh, M.; Singh, S. A Comparison of the Sealing Ability of Various Temporary Restorative Materials to Seal the Access Cavity: An In Vitro Study. *J. Pharm. Bioallied Sci.* **2016**, *8*, 42. [CrossRef]
19. Jensen, A.-L.; Abbott, P.V. Experimental Model: Dye Penetration of Extensive Interim Restorations Used during Endodontic Treatment While under Load in a Multiple Axis Chewing Simulator. *J. Endod.* **2007**, *33*, 1243–1246. [CrossRef]
20. Križnar, I.; Seme, K.; Fidler, A. Bacterial Microleakage of Temporary Filling Materials Used for Endodontic Access Cavity Sealing. *J. Dent. Sci.* **2016**, *11*, 394–400. [CrossRef]
21. Kameyama, A.; Saito, A.; Haruyama, A.; Komada, T.; Sugiyama, S.; Takahashi, T.; Muramatsu, T. Marginal Leakage of Endodontic Temporary Restorative Materials around Access Cavities Prepared with Pre-Endodontic Composite Build-Up: An In Vitro Study. *Materials* **2020**, *13*, 1700. [CrossRef]
22. Prabhakar, A.; Rani, N.S. Comparative Evaluation of Sealing Ability, Water Absorption, and Solubility of Three Temporary Restorative Materials: An In Vitro Study. *Int. J. Clin. Pediatr. Dent.* **2017**, *10*, 136–141. [CrossRef]
23. Pieper, C.M.; Zanchi, C.H.; Rodrigues-Junior, S.A.; Moraes, R.R.; Pontes, L.S.; Bueno, M. Sealing Ability, Water Sorption, Solubility and Toothbrushing Abrasion Resistance of Temporary Filling Materials. *Int. Endod. J.* **2009**, *42*, 893–899. [CrossRef]
24. Ghani, Z.; Mohamad, D.; Hussein, T.; Bakar, W.Z. The Assessment of Surface Roughness and Microleakage of Eroded Tooth-Colored Dental Restorative Materials. *J. Conserv. Dent.* **2014**, *17*, 531. [CrossRef]
25. Jensen, A.-L.; Abbott, P.; Salgado, J.C. Interim and Temporary Restoration of Teeth during Endodontic Treatment. *Aust. Dent. J.* **2007**, *52*, S83–S99. [CrossRef]
26. Ferreira, M.M.; Abrantes, M.; Ferreira, H.; Carrilho, E.V.; Botelho, M.F. Comparison of the Apical Seal on Filled Root Canals with Topseal® vs. MTA Fillapex® Sealers: A Quantitative Scintigraphic Analysis. *Open J. Stomatol.* **2013**, *3*, 128–132. [CrossRef]
27. Carrilho, E.; Abrantes, M.; Paula, A.; Casalta-Lopes, J.; Botelho, M.; Ferreira, M. Microleakage Study of a Restorative Material via Radioisotope Methods. *Rev. Port. Estomatol. Med. Dent. Cir. Maxilofac.* **2014**, *55*, 129–134. [CrossRef]
28. Paranjpe, A.; Jain, S.; Alibhai, K.J.; Wadhwani, C.P.; Darveau, R.P.; Johnson, J.D. In Vitro Microbiologic Evaluation of PTFE and Cotton as Spacer Materials. *Quintessence Int.* **2012**, *43*, 703–707.
29. Gale, M.S.; Darvell, B.W. Thermal Cycling Procedures for Laboratory Testing of Dental Restorations. *J. Dent.* **1999**, *27*, 89–99. [CrossRef]
30. Eliasson, S.T.; Dahl, J.E. Effect of Thermal Cycling on Temperature Changes and Bond Strength in Different Test Specimens. *Biomater. Investig. Dent.* **2020**, *7*, 16–24. [CrossRef]
31. Morresi, A.L.; D'Amario, M.; Capogreco, M.; Gatto, R.; Marzo, G.; D'Arcangelo, C.; Monaco, A. Thermal Cycling for Restorative Materials: Does a Standardized Protocol Exist in Laboratory Testing? A Literature Review. *J. Mech. Behav. Biomed. Mater.* **2014**, *29*, 295–308. [CrossRef]

32. Al Khowaiter, S.; Al-Bounni, R.; Binalrimal, S. Comparison of Dentinal Microleakage in Three Interim Dental Restorations: An In Vitro Study. *J. Int. Soc. Prev. Community Dent.* **2022**, *12*, 590. [CrossRef] [PubMed]
33. Faul, F.; Erdfelder, E.; Buchner, A.; Lang, A.-G. Statistical Power Analyses Using G*Power 3.1: Tests for Correlation and Regression Analyses. *Behav. Res. Methods* **2009**, *41*, 1149–1160. [CrossRef]
34. de Dios Teruel, J.; Alcolea, A.; Hernández, A.; Ruiz, A.J.O. Comparison of Chemical Composition of Enamel and Dentine in Human, Bovine, Porcine and Ovine Teeth. *Arch. Oral Biol.* **2015**, *60*, 768–775. [CrossRef]
35. Altaii, M.; Broberg, M.; Cathro, P.; Richards, L. Standardisation of Sheep Model for Endodontic Regeneration/Revitalisation Research. *Arch. Oral Biol.* **2016**, *65*, 87–94. [CrossRef] [PubMed]
36. Szabelska, A.; Tatara, M.R.; Krupski, W. Morphometric, Densitometric, and Mechanical Properties of Maxillary Teeth in 5-Month-Old Polish Merino Sheep. *Cells Tissues Organs* **2018**, *206*, 196–207. [CrossRef]
37. Anderson, R.W.; Powell, B.J.; Pashley, D.H. Microleakage of Temporary Restorations in Complex Endodontic Access Preparations. *J. Endod.* **1989**, *15*, 526–529. [CrossRef] [PubMed]
38. Gandolfi, M.G.; Taddei, P.; Pondrelli, A.; Zamparini, F.; Prati, C.; Spagnuolo, G. Demineralization, Collagen Modification and Remineralization Degree of Human Dentin after EDTA and Citric Acid Treatments. *Materials* **2018**, *12*, 25. [CrossRef] [PubMed]
39. Jenkins, S.; Kulild, J.; Williams, K.; Lyons, W.; Lee, C. Sealing Ability of Three Materials in the Orifice of Root Canal Systems Obturated with Gutta-Percha. *J. Endod.* **2006**, *32*, 225–227. [CrossRef] [PubMed]
40. Sauáia, T.S.; Gomes, B.P.F.A.; Pinheiro, E.T.; Zaia, A.A.; Ferraz, C.C.R.; Souza-Filho, F.J. Microleakage Evaluation of Intraorifice Sealing Materials in Endodontically Treated Teeth. *Oral Surg. Oral Med. Oral Pathol. Oral Radiol. Endodontol.* **2006**, *102*, 242–246. [CrossRef]
41. Versiani, M.A.; Abi Rached-Junior, F.J.; Kishen, A.; Pécora, J.D.; Silva-Sousa, Y.T.; de Sousa-Neto, M.D. Zinc Oxide Nanoparticles Enhance Physicochemical Characteristics of Grossman Sealer. *J. Endod.* **2016**, *42*, 1804–1810. [CrossRef]
42. Wilson, A.D.; Batchelor, R.F. Zinc Oxide-Eugenol Cements: II. Study of Erosion and Disintegration. *J. Dent. Res.* **1970**, *49*, 593–598. [CrossRef]
43. Sarkar, N.K. Metal–Matrix Interface in Reinforced Glass Ionomers. *Dent. Mater.* **1999**, *15*, 421–425. [CrossRef]
44. Bowen, R.L.; Marjenhoff, W.A. Dental Composites/Glass Ionomers: The Materials. *Adv. Dent. Res.* **1992**, *6*, 44–49. [CrossRef]
45. Thornton, J.B.; Retief, D.H.; Bradley, E.L. Marginal Leakage of Two Glass Ionomer Cements: Ketac-Fil and Ketac-Silver. *Am. J. Dent.* **1988**, *1*, 35–38.
46. Yamauchi, K.; Tsujimoto, A.; Jurado, C.A.; Shimatani, Y.; Nagura, Y.; Takamizawa, T.; Barkmeier, W.W.; Latta, M.A.; Miyazaki, M. Etch-and-Rinse vs Self-Etch Mode for Dentin Bonding Effectiveness of Universal Adhesives. *J. Oral Sci.* **2019**, *61*, 549–553. [CrossRef]
47. Carrilho, E.; Cardoso, M.; Marques Ferreira, M.; Marto, C.; Paula, A.; Coelho, A. 10-MDP Based Dental Adhesives: Adhesive Interface Characterization and Adhesive Stability—A Systematic Review. *Materials* **2019**, *12*, 790. [CrossRef]
48. Jacker-Guhr, S.; Sander, J.; Luehrs, A.-K. How "Universal" Is Adhesion? Shear Bond Strength of Multi-Mode Adhesives to Enamel and Dentin. *J. Adhes. Dent.* **2019**, *21*, 87–95. [CrossRef]
49. Rengo, C.; Goracci, C.; Ametrano, G.; Chieffi, N.; Spagnuolo, G.; Rengo, S.; Ferrari, M. Marginal Leakage of Class V Composite Restorations Assessed Using Microcomputed Tomography and Scanning Electron Microscope. *Oper. Dent.* **2015**, *40*, 440–448. [CrossRef]
50. Pereira, I.R.; Carvalho, C.; Paulo, S.; Martinho, J.P.; Coelho, A.S.; Paula, A.B.; Marto, C.M.; Carrilho, E.; Botelho, M.F.; Abrantes, A.M.; et al. Apical Sealing Ability of Two Calcium Silicate-Based Sealers Using a Radioactive Isotope Method: An In Vitro Apexification Model. *Materials* **2021**, *14*, 6456. [CrossRef]
51. Rengo, C.; Spagnuolo, G.; Ametrano, G.; Juloski, J.; Rengo, S.; Ferrari, M. Micro-Computerized Tomographic Analysis of Premolars Restored with Oval and Circular Posts. *Clin. Oral Investig.* **2014**, *18*, 571–578. [CrossRef]

MDPI AG
Grosspeteranlage 5
4052 Basel
Switzerland
Tel.: +41 61 683 77 34

Journal of Functional Biomaterials Editorial Office
E-mail: jfb@mdpi.com
www.mdpi.com/journal/jfb

www.ingramcontent.com/pod-product-compliance
Lightning Source LLC
Chambersburg PA
CBHW062052220926
51850CB00010B/2739